Engineering and War:

Militarism, Ethics, Institutions, Alternatives

Synthesis Lectures on Engineers, Technology, and Society

Editor
Caroline Baillie, *University of Western Australia*

The mission of this lecture series is to foster an understanding for engineers and scientists on the inclusive nature of their profession. The creation and proliferation of technologies needs to be inclusive as it has effects on all of humankind, regardless of national boundaries, socio-economic status, gender, race and ethnicity, or creed. The lectures will combine expertise in sociology, political economics, philosophy of science, history, engineering, engineering education, participatory research, development studies, sustainability, psychotherapy, policy studies, and epistemology. The lectures will be relevant to all engineers practicing in all parts of the world. Although written for practicing engineers and human resource trainers, it is expected that engineering, science and social science faculty in universities will find these publications an invaluable resource for students in the classroom and for further research. The goal of the series is to provide a platform for the publication of important and sometimes controversial lectures which will encourage discussion, reflection and further understanding.

The series editor will invite authors and encourage experts to recommend authors to write on a wide array of topics, focusing on the cause and effect relationships between engineers and technology, technologies and society and of society on technology and engineers. Topics will include, but are not limited to the following general areas; History of Engineering, Politics and the Engineer, Economics , Social Issues and Ethics, Women in Engineering, Creativity and Innovation, Knowledge Networks, Styles of Organization, Environmental Issues, Appropriate Technology.

Engineering and War: Militarism, Ethics, Institutions, Alternatives
Ethan Blue, Michael Levine, Dean Nieusma
December 2013

Engineers Engaging Community: Water and Energy
Carolyn Oldham, Gregory Crebbin, Stephen Dobbs, Andrea Gaynor
February 2013

Engineering: Women and Leadership
Corri Zoli, Shobha Bhatia, Valerie Davidson, Kelly Rusch
2008

Bridging the Gap Between Engineering and the Global World: A Case Study of the Coconut (Coir) Fiber Industry in Kerala, India
Shobha K. Bhatia, Jennifer L. Smith
2008

Engineering and Social Justice
Donna Riley
2008

Engineering, Poverty, and the Earth
George D. Catalano
2007

Engineers within a Local and Global Society
Caroline Baillie
2006

Globalization, Engineering, and Creativity
John Reader
2006

Engineering Ethics: Peace, Justice, and the Earth
George D. Catalano
2006

Engineering and War: Militarism, Ethics, Institutions, Alternatives
Ethan Blue, Michael Levine, and Dean Nieusma

ISBN: 978-3-031-00985-3 print
ISBN: 978-3-031-02113-8 ebook

DOI 10.1007/978-3-031-02113-8

A Publication in the Springer Nature series
SYNTHESIS LECTURES ON ADVANCES IN AUTOMOTIVE TECHNOLOGY
#20 Series Editor: Caroline Baillie, University of Western Australia

Series ISSN   1933-3633 Print   1933-3641 Electronic

Engineering and War:

Militarism, Ethics, Institutions, Alternatives

Ethan Blue
University of Western Australia, Perth, Australia

Michael Levine
University of Western Australia, Perth, Australia

Dean Nieusma
Rensselaer Polytechnic Institute, Troy, New York, U.S.

SYNTHESIS LECTURES ON ENGINEERS, TECHNOLOGY, AND SOCIETY #20

ABSTRACT

This book investigates the close connections between engineering and war, broadly understood, and the conceptual and structural barriers that face those who would seek to loosen those connections. It shows how military institutions and interests have long influenced engineering education, research, and practice and how they continue to shape the field in the present. The book also provides a generalized framework for responding to these influences useful to students and scholars of engineering, as well as reflective practitioners. The analysis draws on philosophy, history, critical theory, and technology studies to understand the connections between engineering and war and how they shape our very understandings of what engineering is and what it might be. After providing a review of diverse dimensions of engineering itself, the analysis shifts to different dimensions of the connections between engineering and war. First, it considers the ethics of war generally and then explores questions of integrity for engineering practitioners facing career decisions relating to war. Next, it considers the historical rise of the military-industrial-academic complex, especially from World War II to the present. Finally, it considers a range of responses to the militarization of engineering from those who seek to unsettle the status quo. Only by confronting the ethical, historical, and political consequences of engineering for warfare, this book argues, can engineering be sensibly reimagined.

KEYWORDS

engineering reform, engineering profession, history of engineering, epistemology, ethics, just war, integrity, military-industrial complex, non-lethal weapons, peace, social justice

Contents

Preface

This book is the result of a fortunate alignment of interests among three scholars with very distinct research profiles, analytic inclinations, and even conceptual styles. The prospect of the book came through our joint participation in Caroline Baillie's research project, "Engineering Education for Social and Environmental Justice." This project brought together scholars from a wide variety of disciplines—across engineering, the social sciences, and the humanities—with the goal of forging common ground around the project's theme. While the participants' primary goal was the creation of innovative interdisciplinary educational content intended specifically for the engineering classroom, our interdisciplinary collaboration in the larger project resulted in heady conversations, occasional clashing of perspectives, and even some wild ideas for collaborative mini-projects that forced us out of our comfort zones and into spaces of genuine intellectual exploration. This book is one such example.

None of the authors of this book has expertise precisely at the intersection of engineering and war/militarism, yet all of us have expertise that intersects with engineering or militarism. This project required each of us to put on the table what we thought best contributed to answering the question: *How can engineers, engineering students, and engineering educators better understand and respond to the many explicit and subtle forces steering engineering work toward the ends of warfare?* While none of us could provide a convincing answer to that question on our own, together, we thought, a more persuasive response could be formulated. This book is our answer to the question. We acknowledge our answer remains partial, that there are major forces left unaddressed, that other scholars have offered contributions that we have neglected. This is necessarily the case, a condition of both the ambitiousness of the question we sought to answer and the intersections of our own scholarly expertise. But to say that our answer is partial is not to say that it isn't necessary, tackling as it does dimensions of engineering militarism that are rarely treated together.

The writing of this book has not been without its challenges, not least of which were confronting the very disciplinary differences that animated the collaboration. Rather than attempting to wash out those differences, we have chosen a structure for the book that puts them into relief. Each of us has put his mark on the entire manuscript, but individual sections also highlight each of our distinct approaches and styles. Michael Levine, a philosopher, provided the philosophical background on the ethics of war in Chapter 2 as well as the deep analysis of integrity to provide guidance for individual engineers' career decision making in Chapter 3. Ethan Blue, a social and cultural historian, provided the historical analysis of Chapters 4 and 5, which highlights the remarkable influence—both direct and indirect—of the military-industrial-academic complex in steering the work of huge numbers of engineers in the U.S. and beyond over seven decades. Dean Nieusma, whose background is in science and technology studies, provided the material on engineering reformers for Chapter 6. He also constructed most of the scaffold for the book in Chapters 1 and 7, weaving together content and insights from all three authors into an overarching analytic framework.

The challenges of bringing this diverse material together were compensated by the commitment each author brought to the project and, not least, the learning that resulted from disciplinary button-pushing and the need to interrogate one's own assumptions as scholar and would-be reformer of engineering. We hope and expect that our readers will follow a similar journey—sometimes finding useful insights and sometimes experiencing disquiet, sometimes identifying productive tensions and sometimes disagreeing outright, but always challenging the too-easy dismissal of that which diverges from one's own worldview.

Acknowledgments

The authors wish to acknowledge all those who contributed to the completion of this book. First, we acknowledge Caroline Baillie's effort to bring us together as part of her widely multidisciplinary research grant titled, "Engineering Education for Social and Environmental Justice." We acknowledge financial support for that grant (CG10-1519) from the Australian Teaching and Learning Council, which made possible early research for the book. In the context of this larger initiative, Caroline encouraged us to collaborate on this project and to set our collective expectations high in terms of producing a full manuscript for her series, Synthesis Lectures on Engineers, Technology and Society. Caroline provided on-going support for the project, stepping in when necessary to help keep the project moving forward and stepping back when appropriate to allow us to sort out for ourselves the shape of the final project. Caroline, as Series Editor for the book, and *Morgan and Claypool* editor, Andrea Koprowicz, both exercised great patience and provided valuable assistance in moving the book into publication.

We also wish to thank two anonymous reviewers for thoughtful critique and helpful suggestions on a prior version of the manuscript. Many of their insights have been integrated into the book's present version, and while some of their feedback extended beyond the scope of this project, they have helped us to be more explicit in framing our work to make clearer what we have sought to achieve with the book. In addition to these formal reviews, George Catalano and Camar Díaz Torres provided thoughtful criticism and insightful suggestions that have greatly improved this present version. Our thanks also go to Damian Cox and Marguerite La Caze, whose co-authored work with Michael Levine on integrity contributed greatly to Chapter 3. Michael Levine is grateful to the Institute for Advanced Studies at Durham University for a Senior Fellowship in 2013.

Finally, we wish to thank the Engineering, Social Justice, and Peace community for providing an intellectual home for the sort of work carried out here. Deeply interdisciplinary work of this sort is difficult in most academic institutional contexts. The ESJP community provides both refuge and a model for balancing high expectations regarding analytic rigor with great generosity in terms of research approach and scholarly style.

CHAPTER 1

The Close Alignment of Engineering and Warfare

Military values and goals have long played a significant role in shaping engineering, and that influence remains today even though it is often hidden. This book explores the nature of that influence, focusing on its subtler *structural dimensions* as well as important *ethical implications* that arise from such structures. The structural dimensions of influence we will explore span individual values, disciplinary problem-solving protocols and assumptions, organizational missions, and even entire national research enterprises. These structures are patterns of social arrangement that repeat across diverse domains and often impact people, as individuals and as actors embedded within larger organizations, in ways that escape their awareness.

We will focus our attention throughout this book on two such structures: the *conceptual and institutional frameworks* that direct much engineering activity toward the purposes of warfare even when many of the engineers participating in this activity never explicitly decide to pursue careers advancing warfare technologies. *Conceptual frameworks* are the mental models people use to determine the possible, reasonable, and preferred courses of action in any given situation. In a way, they are filters we use to make sense of complex circumstances or environments in order to coordinate our actions with those of others. *Institutional frameworks*, on the other hand, are the logics and rules that operate at the level of social systems—from economic incentive structures to formalized political procedures to informal social norms.

Using these tools, we investigate how military interests and values have become insinuated into engineering in ways that go far beyond individual engineers choosing careers to advance warfare technologies. But those who choose such careers are part of our story, too.

1.1 DIRECT AND INDIRECT CONNECTIONS: ENGINEERING, WARFARE, MILITARISM

While it is difficult to determine the exact number of engineers doing military-related work, Papadopoulos and Hable provide a useful baseline. Using federal employment and research statistics in the U.S., they calculate that "8.8% of professional engineering effort is devoted to defense-related activity [which is] about 3 times higher than for the overall workforce defense effort (2.8%)" (2008, 4). Since this percentage is of "full-time equivalents," and many engineers work only part-time on military projects, "in reality, a much higher percentage of engineers are likely to be engaged in defense-related work as part of their employment" (2008, 5). Papadopoulos and Hable go on to estimate that, for the period 1986 to 2006, "US defense R&D accounted for approximately 57% of all federal R&D expenditures" (2008, 3) and "about 51% of federally funded engineering research is defense related" (6). Ultimately, estimates of the total percentage of U.S. engineers involved in military research range from 30 to 60% (Papadopoulos and Hable 2008; Meiksins 1996).

Determining what constitutes "military" work is even more complex. The interconnections between the military, industry, and universities, in particular, have become so encompassing that even engineering education, practice, and research that may appear to exist apart from military application is implicated either as growing out of military goals and interests or as potentially usable for military purposes. As engineering historian David Noble (1979) has shown, U.S. educational reforms after World War I, particularly in engineering, were designed to leverage the perceived positive educational results of military training, namely to provide disciplined engineers who were ready and willing to serve corporate and military interests. Cornell University Dean of Engineering William Streett similarly reflected on the "intriguing bond between military enterprise, with its emphasis on discipline, loyalty, and uniformity, and other institutions, including science, engineering, government, and education" (1993, 3).

These are some of the explicit connections between engineering and militarism. As our analysis unfolds, we will pay increasing attention to the implicit connections, those that are more difficult to identify and even trickier to characterize. While most of our empirical material focuses on the two extremes—engineers working for war and those working for peace—our analysis will show how that larger group of engineers "in between" these poles is also implicated. To do this, we will highlight how the subtler connections between engineering and militarism implicate a much broader community of engineers in advancing militaristic values and goals, even though they do so indirectly and unintentionally.

Given the intertwined nature of engineering and militarism, it can be difficult for engineers who have a clear commitment against participating in war-related activities to manage their careers in a way that honors their values. For others who have no particular commitment against engineering for warfare, considerable incentives exist to pull them in that direction. Drawing on traditions in history, technology studies, critical theory, and the philosophy of ethics, this book explores the nature of the connections between engineering and militarism, focusing particularly on the conceptual and institutional frameworks that make the alignments between the two appear to be "natural," if not inevitable. While the predominant thrust of this book is conceptual—seeking to better understand and characterize subtle connections and alignments—we hope the conceptual contributions produced can be applied, reflectively if not didactically, in the practical settings in which engineers, engineering students, and engineering educators find themselves.

This chapter begins our investigation into the intersections of engineering, warfare, and militarism by reviewing some of the ways engineering is defined, starting with the birth of the term "engineer" and then working through several different definitions that are in circulation today. Understanding the comfortable fit between engineering and militarism requires first looking at what engineering is understood to be.

1.2   WHAT IS ENGINEERING?

Many books on the topic of engineering start with definitions, as does this one. But our purpose in reviewing definitions of engineering is less to formulate the "correct" definition and more to identify a *range* of definitions as well as the various *dimensions* of engineering that those definitions convey. Before looking at some of the specific characteristics of engineering that facilitate its alignment with militarism, we will show how different definitions of engineering enable different types of analysis into the nature of engineering.

The term "engineer" dates from the 14th century and explicitly ties engineering to warfare:

Engineer (origin): Middle English (denoting a designer and constructor of fortifications and weapons; formerly also as ingineer). (O.E.D.)

...early 14c., "constructor of military engines," from O.Fr. engigneor... sense of "inventor, designer" is recorded from early 15c.; civil sense, in ref. to public works, is recorded from c.1600 (http://dictionary.reference.com/browse/engineer)

Insofar as one understands "engineering" to entail the design and making of technology, it can be connected to military purposes even earlier. Egyptian, Mayan, Greek, Roman, Aztec, and other ancient civilizations were built, in no small part, upon technologies of fortification and weaponry. Surely, the history of technologies of warfare is even older, likely dating from the very beginnings of civilization, or at least from the time human or even pre-human species began to make and use tools of any sort.

In contemporary definitions of engineering, direct connections to technologies of warfare have been lost. What remains in these definitions illustrates the widely varying terrain of what "engineering" is understood to be. While some common definitions may seem reasonable as far as they go, they often hide as much as they explain. For example, consider this historical but frequently used definition:

Engineering is the application of science to the common purpose of life.[1]

Here we see two widely identified facets of engineering. Its method is the application of science, and its *domain of application*, in contemporary language, is everyday life. Both of these seem reasonable enough. Engineers do, in fact, apply scientific principles to address problems of everyday life. Yet this definition hardly seems sufficient to capture how engineers are distinct from many other professionals—applied scientists, doctors, organizational theorists, or even market researchers.

[1] Widely attributed to Count Rumford, 1799.

A more contemporary definition expands on both the methods employed by engineering and its domains of application:

Engineering: The application of scientific and mathematical principles to practical ends such as the design, manufacture, and operation of efficient and economical structures, machines, processes, and systems.[2]

Here, *method* is expanded to include the application of mathematical principles alongside science. Similarly, "practical ends" are elaborated to include common categories of engineering practice: design, manufacture, and operation. Further, the *domains of application* of those practices are specified according to categories of engineered technologies: structures, machines, processes, and systems.

In this definition, the lines between engineers and other professionals who draw upon science are more clearly drawn. The ways in which engineers "apply science and mathematical principles"— that is, through design, manufacture, and operation of technologies—are identified, but so too are the overarching goals: making technology efficient and economical. As a result, this definition appears to be more robust than the prior one, in part because it is more specific in identifying common methods, practices, applications, and goals of engineering.

But this definition also contains important ambiguities. For example, "efficiency," in generic terms, means doing more with less. At first glance, this seems to be both an obvious and a universal good. Attempting to apply the principle in engineering practice, however, requires determining what, exactly, it is that we want more of and by using less of what else. More energy out of the same size solar panel certainly seems desirable, but at what cost in terms of financial, material, and labor resources? What cost in terms of environmental impact of production? Disruptions to global supply chains in the fragile solar market? Reparability and serviceability in poor, remote regions? Ultimately, efficiency is calculated as a *ratio*, and it matters what the numerator and denominator are.

Making engineered systems "economical" is another near-universal, but generic goal. Economical is simply another type of efficiency goal, focusing specifically on financial resource efficiencies, and the same questions apply: Cost savings may sometimes be achieved through technical refinements exclusively, with limited upstream financial pressures. More commonly, however, cost savings are achieved by tightening margins at each stage of the production process, which is often financially disadvantageous to upstream laborers—miners, factory workers, transportation workers, etc. Indeed, following Noble (1979), one of the driving forces behind technological innovation since the industrial revolution has been factory owners' effort to cut labor costs or else to lessen the bargaining power of unionized workers.

The economics concept of *externalities* refers to costs occurring as a result of a transaction that are not reflected in its price. For example, the pollution created by a person's car is not reflected

---

[2] The Free Dictionary. Accessed 19 November 2012 at http://www.thefreedictionary.com/engineering.

in the price of the car or its gasoline. The "cost" of pollution is borne not by the individual driver directly but by society at large, including future generations, through negative health impacts, environmental decline, lowered quality of life, etc. Most economic transactions entail externalities of some sort—some immediate and others distant in time and geography.

The same type of analysis can be applied to any type of efficiency improvement, such as a higher-efficiency solar panel that, say, relies more extensively on rare earth elements. Whereas the high financial cost of these materials may be included in the price, the global depletion of such strategically important yet finite materials is not. Similarly, in today's world, significant technical and economic efficiencies have been achieved through massive reliance on petroleum derivatives in multiple domains of human life, but with equally significant externalized costs in terms of environmental degradation, geopolitical instability, and long-term energy insecurity.

Despite the ambiguities, we suspect most engineers would feel comfortable with a definition of engineering that includes the goals of efficient and economical technology development. Certainly, they are desirable goals, the nuances and trade-offs described above notwithstanding. In fact, we agree. Yet, we raise the question of terminology to make a point. An important part of understanding any social phenomena in the world—including understanding what engineering is in people's minds and actions—is identifying the assumptions that underlie and motivate them. And part of what makes engineering so intriguing is the extent to which goals such as "efficient and economical" output can be included in its definition without raising eyebrows.

Part of what makes these particular characteristics so intriguing is that they help to establish boundaries around what is and is not "engineering" when it comes to technology development. While we agree with the sensibility of striving for efficiency and economical technologies, we posit that they are patently "social" concerns that exist beyond what might be considered the "technical core" of engineering expertise. Systematic exploration of such concerns requires following paths that soon diverge from the technology at hand, as indicated in the analysis above. Quickly, one is stepping into fields traditionally defined as sociology, history, philosophy—the disciplines that have more commonly dedicated themselves to exploring entire economic systems, forms of social organization, international trade relations, even capitalism itself. Such analyses do not seem to fit comfortably with ordinary understandings of what engineering is or what engineers are expected to know how to do.

Let's look at one more definition to further explore how boundary setting around engineering can help us better understand the terrain upon which our analysis will unfold:

> Engineering is the science, skill, and profession of acquiring and applying scientific, economic, social, and practical knowledge, in order to design and also build structures, machines, devices, systems, materials and processes.[3]

---

[3] Wikipedia. The Free Encyclopedia. First sentence of entry on "Engineering." Accessed 17 November 2012 at http://en.wikipedia.org/wiki/Engineering.

Compared to the prior definition, a little is lost—mathematical principles, practical ends, and operation of technologies—but a lot is gained: skill, the profession, acquiring knowledge, social and practical dimensions of engineering knowledge, and materials as an engineering output. We have entered terrain far more contoured than "the application of science."

In the most general terms, this definition points to what engineers *know*, what they *do*, and how they *relate* to one another and society at large (i.e., as a "profession"). Each of these dimensions can help us understand the nature of the alignment between engineering and militarism, and, hence, each will be elaborated in turn.

1.2.1   ENGINEERING AS A DOMAIN OF KNOWLEDGE

Understanding what engineering is obviously requires engaging the types of knowledge that constitute it (Vincenti, 1990). *Epistemology* is a term scholars—from philosophers to anthropologists—use to identify a specific way of knowing the world. There is not a single way of understanding the world, but many; therefore, there are many epistemologies. In each of the definitions above, we see different facets of what might be called an engineering epistemology: first as one founded in science, then as founded in science plus mathematics, and then in scientific, economic, social, and practical (see, e.g., Ferguson 1992) domains. The epistemology identified in the last definition is noteworthy because it adds "social" knowledge domains alongside the more typical scientific, economic, and practical ones, which is particularly relevant to the analysis that this book embarks on.

The addition of "social knowledge" to a definition of engineering may appear to be somewhat paradoxical. After all, stereotypes play on the extent to which social expertise—from interpersonal communication skills to social analytic abilities—is *absent* among engineers. The preceding consideration of efficiency and economic externalities also supports an assessment of engineering knowledge that sticks close to the "technical core" of technology development. Nevertheless, a wide range of social knowledge is clearly required for successful engineering practice in any context, from communicating effectively with clients to understanding how to navigate large, bureaucratic organizations to working collaboratively on multidisciplinary teams.

There is a long and rich history of debate surrounding which types of social knowledge should be included as *constitutive of engineering* and which should be excluded as potentially important but not really "engineering knowledge" (Pawley, 2012; Christensen et al., 2009; Law, 1987). These debates shed some light on how engineering has come to be understood as it is in the present, namely, *technical at its core* with "social knowledge" relegated to the *context* of engineering work and not the work itself.

Here and throughout this book, we will challenge this simple distinction—between the technical core and social context of engineering work—to show how military influences on engineering are not limited to one particular application of engineering knowledge, but are built into the very assumption of a "technical core" at the heart of an engineering epistemology.

1.2.2 ENGINEERING AS A SET OF PRACTICES

Beyond the familiar if abstract domain of disciplinary knowledge lays the wide range of activities engineers engage in while doing engineering work—what we will call *engineering practice*. Whereas a philosopher might start with epistemology in answering with the question, "What is engineering?," a sociologist is likely to start instead with practice: "Engineering is as engineers do." Of course, what engineers do is conditioned by what they know (and what they do not know!) as well as the other dimensions of engineering that we will discuss below. Furthermore, starting with those who are *recognized as engineers* necessarily limits consideration to engineering "insiders," and ignores those who are arguably doing engineering work but are not so recognized (Cockburn and Ormrod, 1993). Still, looking at prototypical engineering practices helps to shed light on the nature of the connections between engineering and militarism.

Engineering practice has been studied from a variety of perspectives to illuminate different facets of engineers' work, including the gendered dimensions of engineering practice (Tonso, 2007, Cockburn and Ormrod, 1993), the everyday activities of wide-ranging practicing engineers (Bucciarelli, 1994; Vinck, 2003), the particular organizational contexts in which engineers work (Kunda, 1992), and the broad employment patterns evident by looking across the entire field of engineering. Some of the broad patterns identified by looking across the field include its close alignment with state and corporate priorities, including nationalist expansion efforts (Lucena, 2005), obedience training through rigid hierarchy and regimentation of daily life (Hacker, 1989), control over labor (Noble, 1979), and a tendency toward highly fragmented and compartmentalized work (Meiksins, 1996, 83) that insulates narrow technical problem solving from concerns over systems integration and technology application.

The close alignment between corporate and state interests and engineering is created most directly through patronship, where corporate and state organizations employ engineers to work toward corporate or state goals. Simply put, these interests provide the financial resources and set the agenda for engineering projects. Because they employ such a large percentage of engineers, as described above, they also shape the nature of the field as a whole. Especially in the U.S. post-World War II, the "strong relationship between engineers and the state-controlled defense sector" both contributed to "a rapid growth in the numbers of engineers" and "increased the percentage of engineers who were directly or indirectly involved in government and/or defense-related work" (Meiksins, 1996, 81–82).

Apart from defense-related work specifically, a major theme across studies of engineering practice is the extent to which engineers are employed by large, bureaucratic firms, regardless of sector. As of 2008, over 60% of scientists and engineers in the U.S. were employed by organizations with more than 100 employees and nearly 20% by organizations with over 25,000 employees (NSF, http://www.nsf.gov/statistics/seind12/pdf/c03.pdf).

While the number of engineers—in total and in military-related work—grew rapidly in the U.S. post-World War II, military perspectives have informed engineering since its creation and have influenced engineering practices far beyond the character of employment patterns. Among the work practices that define engineering and distinguish it from other professional groups are its comfortable integration within hierarchical organizations, its heavy reliance on command-and-control problem solving, a high degree of division of labor and expertise within the field, and its notably masculinist culture. As Noble's (1979) history of engineering shows, these attributes are not coincidental, nor are they shaped exclusively by postwar employment patterns, but instead they were designed into the field from its beginning and as it evolved alongside the growth of corporate capitalism.

1.2.3   ENGINEERING AS A PROFESSION

Related to employment practices and the nature of engineering work—as well as questions around what knowledge is appropriate to the field—is the status of engineering *as a profession*. The terminology surrounding the status of different occupational groups is often used loosely, including the use of the word "profession." As with the definition of engineering itself, we do not seek to promote or enforce any particular definition of profession or to weigh in on the question over whether engineering should properly be understood as a profession. Instead, we identify *the question over engineering's status as a profession itself* as constitutive of the field of engineering over its history and in the present.

Casual use of the word "profession" in describing engineering often simply refers to engineering as an occupational group—people identify as engineers, are identified by others as engineers, and are often formally employed with the title "engineer." Increasingly, these people are graduates of educational programs in various recognized sub-disciplines within engineering, many of such programs being accredited by national or international accrediting bodies (such as ABET[4] and others). The extent to which engineers are certified by standing professional bodies varies considerably by industry and national context (see, e.g., Meiksins & Smith, 1996), but in the U.S. only roughly 20% of engineering program graduates are certified as Professional Engineers by the National Society of Professional Engineers.

A more formal use of the word "profession" invokes the degree to which the occupational group has the autonomy to self-regulate, including in setting educational or training requirements, certifying educational institutions and licensing individual practitioners, and generally constraining the nature of work employed engineers are allowed to do using codes of ethics and the like. There

---

[4] ABET—formerly the Accreditation Board for Engineering and Technology—is the dominant accreditation body for engineering and related programs in the U.S., and it has increased its reach internationally over the past decades, particularly in Latin America.

is a rich history of efforts to *advance professionalization* within engineering in diverse settings and with diverse motivations.

Because of engineering's historical close alignment with corporate and state/military interests, the field has never achieved the same degree of self-regulation as the prototypical professions of medicine and law. Contrasted with these occupational groups, there is also a stronger historic division within engineering along disciplinary boundaries, so disciplinary professional societies' membership exists somewhat in tension with occupation-wide agenda setting. Engineering, for instance, does not have an organizational equivalent of the American Medical Association. While the National Society of Professional Engineers aspires to represent engineering as a profession, this organization has neither the membership nor the influence of the American Medical Association.

Advocates for strengthening engineering's professional standing commonly identify two distinct goals. First, professionalization promises to improve working conditions and benefits for engineers. Second, and perhaps a loftier goal, professionalization offers some assurances that engineering—as a whole—will be able to contribute more effectively to "the social good" than atomized individual employees. Edwin Layton's *The Revolt of the Engineers* (1986) documents early-20th-century professionalization efforts in the U.S. that were largely motivated by a desire to elevate "social responsibility" within the field. "While certainly not a principled critic of capitalism, Layton famously argued that engineers might serve in 'loyal opposition' to corporate interests" (Nieusma and Blue, 2012).

1.2.4   ENGINEERING AS AN IDEOLOGY

Not fully captured in the three dimensions of engineering identified above, yet essential for understanding engineering as a social phenomenon, is the set of values and the belief systems that are widely shared across engineering sub-communities. We refer to this domain as the *ideology* of engineering, and we seek here to identify some threads of such an ideology that have circulated widely in scholarship on engineering. Of course, as with epistemologies, there is not a single ideology of engineering that is shared equally by all engineers. Nevertheless, components of a broadly shared *dominant ideology*—and the existence of competing but marginalized alternative ideologies—can be identified across the field. Here we characterize this dominant ideology of engineering, even as we recognize its contingent nature and seek to contribute to its erosion.

In her influential book, *Engineering and Social Justice* (2008), Donna Riley explores the terrain of ideologies—what she calls "worldviews" and the "engineering mindsets" that accompany them. She describes five engineering mindsets:

- A desire to help and the persistence to follow through on that desire

- The centrality of military and corporate organizations to engineering practice

- A narrow technical focus (and weakness in other areas)

- Positivism and the myth of objectivity

- Uncritical acceptance of authority

We have alluded to manifestations of most of these above, but will elaborate on Riley's third and fourth dimensions: the narrow technical focus and positivism/myth of objectivity. *Positivism* contends that all rational assertions (i.e., "truths") can be verified by empirical scientific means and/or logical or mathematical proof. In other words, science equals truth, and claims that cannot be scientifically verified should not have legitimate standing. The myth of objectivity suggests that, because (technical) engineering work rests on science, and because science rests on truths, engineering—when done properly—is politically neutral. Engineering, like science, is taken to speak truth to (political) power, or, at least, to operate on another dimension of life that is considered unrelated to politics.[5]

Despite their important role in shaping the dominant ideology of engineering, terms like "positivism" are not often used by engineers and do not always translate well to the ways they understand their own values and beliefs. Instead, as suggested above, engineers understand engineering in terms of its foundations in science and math, its goals of advancing technology and improving efficiency, and its parallel service to both business interests and a broader social welfare. Certainly, for most engineers (and many others), technology innovation in itself advances social welfare. Technology advance *is progress*, a storyline complicated only by illegitimate or inappropriate "applications" of the technology but not attributable to "the technology itself."

The decoupling of "the technology" from its "application" is an important foundation for the ideology of engineering and is central to engineers' myth of objectivity. In this understanding, "technology" can refer to an underlying *idea* or set of *abstracted components* for solving a particular sort of problem as well as a particular material instantiation of that idea (i.e., one particular application). For example, no one car or gun captures the totality of cars or guns *as technologies*. In other words, an actual car in the world is just one instantiation of car technology and captures neither *all* that is car technology nor the *essence* of that technology. At the same time, each instantiation of a car or gun entails lots of additional elements or features that are not constitutive of the underlying technology. Hence, a particular car and a particular gun are both less and more than their underlying technologies in engineers' expansive and sometimes conflicting understanding of the term.

At its core, this assumption captures an organizational reality of engineering employment: engineers are typically hired and paid to make technologies "work," not to decide or even weigh in on how those technologies will be applied—by end users, their employers, their clients, their

---

[5] Granting positivist assumptions, however, would not translate to "neutrality" the way many engineers profess it. Even if it were possible, political neutrality would de facto support the status quo, which itself is a non-neutral political position.

sponsors, or their patrons. However, an important contradiction is bubbling under the surface here. The narrative that engineers design and enable the production of abstract technologies that are somehow independent of their real-world applications is fraught for many reasons, as the following chapters will elaborate. This rejection of empirically verifiable conditions also verges on the unscientific, but to see this requires moving from the physical and into the social sciences. This move fits uneasily—but, we hope, productively—with the dominant ideology of engineering as "neutral."

Social-Technical Dualism

Foundational to every dimension of the dominant ideology of engineering is a dualism between *the technical* and *the social*, where the two domains are understood to be independent from one another if not mutually exclusive. Insofar as engineering is defined as constituted by its "technical core"— the underlying (physical but not social) scientific and mathematical principles—and not by *how*, *when*, and *where* such principles are deployed in the creation of specific technologies embedded in the real (social and physical) world, engineers can insulate themselves from the consequences of their work.

According to this model, engineers are to be held accountable only if their technical work is deficient. Outcomes attributable to social domains are neatly outsourced from engineering, with users, clients, or "politics" usually having responsibility for deleterious results. (Note here the curious exception of a given technology's economic dimensions, as described above; profitability is a patently social concern but fitting comfortably within the engineer's domain.) As will be clear through the book, we argue that a richer knowledge and practice of engineering must draw on the social sciences and the humanities along with the physical sciences and mathematics. Ultimately, this integration demands reconsideration of the social-technical dualism that serves as foundation to the dominant engineering ideology.

Masculinist Values

Another important dimension of the dominant engineering ideology is its masculinist underpinning. Describing engineering as masculinist, as opposed to merely masculine or even male-dominated, serves to highlight the purported or assumed *superiority* of masculine attributes or characteristics—at least within the field. To be sure, engineering is and historically has been dominated by men. Even as participation by women in engineering has trended upward in recent history, the percentage of women in engineering remains far below 50, and the percentage drops precipitously as rank increases. While there has been considerable attention to—and consternation over—the underrepresentation of women (and other groups) in engineering, the problem endures, and for a variety of reasons that includes masculinist values and the exclusion and hostility they create for women.

The role of masculinist values in the dominant engineering ideology is not limited to its negative impact on women's participation, and in fact shapes the experience of all engineers, male and female (Cockburn and Ormrod, 1993). In fact, it is an important component of gatekeeping within the field—the process of determining who is and who is not allowed to become an engineer. Arguably, the very presence of a "weed-out" culture in engineering (Seymour and Hewitt, 1997)—where extensive and deliberately punitive testing predominates—derives from a masculinist value system that valorizes an individualistic, "sink or swim" approach to learning, where students must prove themselves worthy, not only to their professors but also to their peers and even themselves as well.

The parallels between engineering and military institutions concerning masculinst values has not been lost on feminist scholars of engineering and technology, even as it remains under recognized in broader culture. As feminist theorist Sally Hacker puts it, "Although many other forces mold both gender and technology, the effects of military institutions are as important as they are ignored" (1989, 60). Hacker goes on to describe some of the ways that military institutions—and their particular assumptions around gender, both of masculinity and femininity—provided models for engineering and engineering education. Going back to the earliest engineering schools in Europe, she argues that "Curricula fused technical training with cultural socialization that stressed hierarchy, discipline, loyalty, and self-control to a male-only student body" (1989, 61). With the exception of the qualification "male-only," this quotation arguably remains accurate in describing engineering education to this day.

While many opponents of reform to engineering are likely to dismiss the entire idea of a dominant engineering ideology at play in the field, such hard-nosed posturing is precisely indicative of that ideology in practice: Engineering is hard; toughen up, suck it up, and get back to work. Identifying and exploring this underlying value system are necessary steps in confronting the intimate intertwining of engineering, militarism, and warfare.

1.3   ENGINEERING'S SURPRISING SILENCE ON WARFARE

Given the kinds of close, substantial, and pervasive connections between engineering and the military identified in our opening, one might expect considerable attention paid to questions of warfare within engineering, or at least within engineering educational settings. This is generally not the case. As Vesilind avers, "The effect of military research at engineering schools has largely been ignored by the disciplines that study engineering and engineering education…[despite its key role in] the greatest single decision engineers have to make—whether or not to work in the armaments industry" (Vesilind, 2010, 151). It is debatable whether such a broad claim is analytically justifiable, but clearly Vesilind believes the decision deserves considerably more attention than it is given at present, not least because the relationship between engineering and warfare goes largely unaddressed within engineering education.

Where coverage of engineering and warfare is most likely to exist in engineering education—within engineering ethics textbooks—the theme is sometimes included but rarely systematically explored. Issues of sustainability, whistleblowing, liability, risk assessment, ethical codes of conduct, global ethics, and even matters of social justice, human welfare, and human rights are all elaborated in the engineering ethics literature; however, war in relation to engineering is often referenced in passing but infrequently elaborated. For example, in their introductory text, *Ethics, Technology and Engineering: An Introduction*, Van de Poel and Royakkers say:

> [E]ngineering is not just about better understanding the world, but also about changing it.… There is an increasing attention to ethics in the engineering curriculum. Engineers are supposed to carry out their work competently and skillfully but also be aware of the broader ethical and social implications of engineering and to be able to reflect on these. (2011, 1)

And yet, war is mentioned neither in their table of contents nor the index, and questions of warfare and militarism are not otherwise systematically addressed.[6]

In *Engineering Ethics: Peace, Justice, and the Earth*, Catalano identifies several pertinent issues surrounding engineering and warfare—lack of attention to war in ethics codes, the influence of militarism on engineering, and weapons systems—but then focuses his attention on the environment and ecology, sustainability, and the like (2006; 62). He does not systematically explore the many questions surrounding warfare in engineering. In *Ethics and Professionalism in Engineering*, McCuen and Gilroy (2010) also discuss engineering ethics in relation to sustainability and the environment, and layer in treatment of climate change, cyber-ethics, food, leadership, business, and risk among many other topics, but again largely skip over challenges surrounding warfare and militarism.

Luckily, there are also some exceptions to the trend. Vesilind's (2010) *Engineering Peace and Justice*, quoted above (and elaborated in Chapter 6), is one of the few recent texts on engineering ethics that systematically addresses warfare, with a chapter on "Military Engineering" as well as an additional section on "War," but this book is not widely circulated and remains infrequently cited. Similarly, W. Richard Bowen's (2009) *Engineering Ethics: Outline of an Aspirational Approach* explicitly covers militarism and warfare, with sections on military technology and, in more detail, cluster munitions, and another section providing a humanistic reflection on the violence of war. Unfortunately, neither Vesilind nor Bowen open up the wider-ranging and subtler alignments between engineering and war, thereby risking to underplay its significance even as they promote paying more attention to this area.

One book in the domain of engineering ethics that more systematically tackles the wide-ranging influence of militarism on engineering is Donna Riley's *Engineering and Social Justice* (2008). In this book, Riley squarely addresses the pervasive role of military interests in engineering,

---

[6] Van de Poel and Royakkers (2011, 209–210) do include a short discussion of military robots and reference the ethical challenges that arise in their deployment.

citing it as part of one of five prevalent mindsets in engineering, as noted above (and also elaborated in Chapter 6). She dedicates a section to militarism's influence on engineering, exploring both its structural dimensions and how it shapes "engineering culture." She also provides a case study of engineering activism against campus military research. As do many engineering ethics scholars who identify militarism as a challenge for engineering, Riley shares our concern with the taken-for-granted status of the relationship and the paucity of systematic reflection regarding its roles and influence.

1.3.1   WHY THE SILENCE?

It may be impossible to identify the source of engineering's broader silence on warfare, but the matter is important enough that some exploration is in order. Here we consider a range of conceptual and institutional structures that are likely to make the connections between engineering and militarism particularly murky for engineers and others.

Psychological Repression

One possible source of silence, admittedly speculative, is what we might identify as a type of *disciplinary repression*. As generations of psychological researchers have made clear, people tend to deflect, self-deceive, or repress those things about themselves that they are unable to rationally and consciously deal with in order to minimize cognitive dissonance and protect their self-images (and egos). That which is repressed—beliefs (whether true or not), desires, and so on—must be important enough to challenge self-conceptions or there would be no need to repress them (see, e.g., Freud, 1919). On this view, the reason for the silence on war in engineering—and in ethics texts in particular—may not be because it is regarded as a small or insignificant problem, but precisely because it is so important. It may well be recognized as an issue that pervades engineering practice in a way that worries and challenges engineers about the ethical implications of their work. Indeed, it may even challenge engineers' conceptions of their own self-worth with regard to what they do.

Political Convictions

Another possible explanation for the relative silence on warfare in engineering is that engineers as an occupational group tend to be *politically conservative* and are more likely to support hawkish approaches to global politics. Existing research identifies engineering as among the most politically conservative of disciplines. Spaulding and Turner (1968) laid the groundwork for such analyses. They showed that engineers tended to be the most politically conservative among college professors in nine academic specialties: botany, engineering, geology, history, mathematics, philosophy, political science, psychology, and sociology. Their study controlled for a range of environmental factors, including parents' political orientations, and concluded "an important element in explaining the differences is the degree of orientation toward social criticism or the application of knowledge in the business world" (1968, 262), where engineering fell far along the "business world" extreme of the spectrum.

A more recent 1984 study by the Carnegie Foundation had similar findings. It showed that engineers were more politically conservative than any other disciplinary grouping, including natural sciences, medicine, economics and business, law, arts and humanities, and social sciences (see Figure 1.1). In this study, engineers self-reported the highest percentage of strongly conservative (15.7%) and moderately conservative (41.9%) political views as well as the lowest percentage of leftist political views (1.4%). Engineers self-reported being liberal (i.e., moderately leftist) at a rate of 20.3%, with only economics and business academics self-reporting a lower percentage in this category at 15.3%.

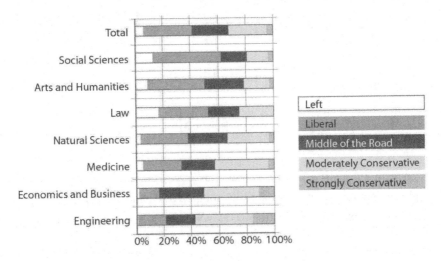

Figure 1.1: Percentage distributions of self-reported political views of U.S. academics by highest degree achieved, males only. (Adapted from: Gambetta and Hertog, 2009 elaborating data from the Carnegie Foundation National Survey of Higher Education, 1984)

An even more recent study found that academic engineers were less conservative than found in the prior studies, but still amongst the most conservative of academic disciplines. Rothman et al. (2005) argue that the academy is overwhelmingly left-leaning relative to other large institutional sectors, but then go on to rate engineering less liberal than 20 other disciplines, with only business ranked as more conservative (see Figure 1.2).

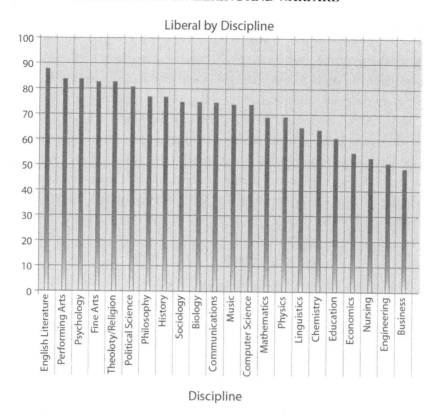

Figure 1.2: Engineering ranks second-least liberal among 22 disciplines. (Adapted from: http://blogs. discovermagazine.com/gnxp/2006/06/math-makes-you-more-conservative/. Accessed June 25, 2013.)

Some authors have gone even further in characterizing the political orientation of engineering, suggesting engineers are over-represented among political and religious fundamentalists (see Bergen and Pandey, 2006). Gambetta and Hertog studied wide-ranging data around Islamic radicalization in order to identify trends surrounding educational type. Careful not to misrepresent the extent of radicalization within engineering as a field, Gambetta and Hertog find: "the number of militant engineers relative to the total population of engineers is miniscule—yet engineers, relative to other graduates, are overrepresented among violent Islamic radicals by three to four times the size we would expect" (2009, 212).

They go on to consider why this might be the case, first dismissing two hypotheses—the structure of existing terrorist networks and that engineers are sought for their technical skills—and then providing two hypotheses they believe are more likely to be the cause—"engineers' peculiar cognitive traits and dispositions [i.e., mindsets], and the special social difficulties faced by engineers in Islamic countries" (Gambetta and Hertog, 2009, 213). The dissemination of these claims, including by the popular press (see, e.g., Popper, 2009), produced considerable skepticism within engineering communities, even though much of the resistance to Gambetta and Hertog's research was directed at claims they do not make, specifically regarding the extent to which violent radicals are present among engineers.

Patronage Structures

Existing somewhere between psychological repression and political conviction is implicit acknowledgement of the role war plays in undergirding engineering's *patronage structures*—who pays for engineering work and in order to achieve what ends. The modern Westphalian world political system that came into being after 1648 saw the slow replacement of European dynastic empires with emergent sovereign nation-states. In this modern system, nation-states tend to see one another as competitors for territory and resources (Held, 1996). Because nation-states are the largest funders of engineering research, engineers' goals and politics become aligned with those of the state (Roland, 2001).

In *Defending the Nation*, Juan Lucena (2005) provides a compelling cultural history of the alignment between engineering and the fortification of the U.S. nation-state since the Sputnik era. Lucena's work focuses on U.S. policy making aimed at boosting the number of engineers and scientists, highlighting the extent to which the state has intervened in incentivizing their education, and hence the development of these occupational groups. Such policies exist prior to and apart from the flow of engineers into lucrative military research positions, which are also sponsored by the state, both directly and indirectly.

We will elaborate in detail the connections among state, corporate, and academic institutions in the creation and employment of engineers in Chapters 4 and 5. For now, we will put the issues of patronage structures in the simplest terms: engineers, like many other occupational groups, are reluctant to bite the hands that feed them.

Dominant Ideologies

Another explanatory approach to understand engineering's relative silence on war is to start from engineers' own conceptions of engineering. Extending from engineering's dominant ideology—particularly its commitment to neutrality and objectivity—provides a different sort of explanation for the silence. For many engineers, "politics" is a word used with derision, often referring not merely to questions about the distribution of power in society or formal political decision-making processes but also to wholly *irrational* and usually *illegitimate* decision-making processes. In this sense, politics is almost a dirty word. Such thinking results in many engineers who could be characterized not merely as *a*political but as *anti*-political—aspiring to a world without political power. To an important degree, such an anti-political orientation is venerated within engineering culture. This position, along with the technical-social distinction that enables it, divorces engineering (read, technical) matters from the "social" issues of militarism and warfare.

When we combine skepticism about political decision making with the assumed neutrality of technology, engineers' lack of professional autonomy, and a high degree of participation on (inherently political) military projects, then it is not surprising that many engineers have difficulty responding to the deep-seated contradictions surrounding the field. Hence, perhaps the most sensible if least evocative explanation for the lack of explicit and systematic critical reflection on militarism in engineering is that the connections are so pervasive, the issues so broad and entwined, that they remain somewhat hidden. They cannot be considered as just one topic among many in engineering ethics, but require a separate and distinct kind of ethical consideration that interrogates the much broader question of what engineering is understood to be.

We intend this book to provide an opportunity for such consideration, at least for those engineers, students, and educators who wish to confront the challenge in their own lives and for their own work. Given the general silence about war in engineering, we seek to stimulate more systematic reflection on this important area. While we will not advise how individual engineers should come down on the issue of whether to participate in the making of military technologies, we do hope to provide a set of questions, historical perspectives, and theoretical insights that will help readers make their own decisions in a deliberate and informed way. And more to the point of our approach, we hope to highlight how disentangling engineering from militarism and war requires sustained, broad-based effort and cannot be achieved solely by variously motivated individuals opting out of military projects.

CHAPTER 2

The Ethics of War

We now shift our attention from a focus on engineering to a focus on war. This chapter elaborates why engineers and engineering educators ought to engage questions of warfare, and then provides some theoretical underpinnings for thinking about war. We begin by highlighting some of the ethical responsibilities faced specifically by engineers, building our analysis upon Paul Taylor's definition of ethics:

> Ethics may be defined as philosophical inquiry into the nature and grounds of morality. The term "morality" is here used as a general name for moral judgments, standards, and rules of conduct. These include not only the actual judgments, standards, and rules to be found in the moral codes of existing societies, but also what may be called ideal judgments, standards, and rules: those which can be justified on rational grounds. Indeed, one of the chief goals of ethics is to see if rational grounds can be given in support of any moral judgments, standards, or rules, and if so, to specify what those grounds are. (Taylor, 1975, 1)

Depending on how engineering—or what it means to be an engineer (see Baillie and Levine, 2013)—is conceived, some ethical issues may be specific to engineering generally, while others arise in connection with particular engineering practices or domains of application. Denying or avoiding the fact that some ethical issues are intrinsic to the practice of engineering is simply another way of making ethical decisions—albeit by default. While we focus our attention here on the responsibilities of engineers *as engineers*, we recognize that engineers also have the same normative moral requirements, duties, and obligations that all people have, both in relation to their conduct in daily life as well in relation to matters of war.

After considering the ethical responsibilities of engineers as engineers, we look at philosophical treatments of the ethics of war by providing an introduction to *just war theory*. This framework helps clarify different dimensions of warfare that are particularly relevant, sometimes even intrinsic, to engineering. In so doing, it helps identify the ethical dilemmas engineers face. In virtue of their engineering skills and capacities, and because of how those skills may well be employed, engineers ought to confront—ideally in an explicit and robust manner—ethical questions that those further removed from technologies of warfare either do not face or at least do not face so directly.

2.1   DO ENGINEERS HAVE SPECIAL ETHICAL RESPONSIBILITIES AROUND WARFARE?

Engineers might well ask why they should have interest in the ethics of war or, even, what it has to do with engineering. They are, after all, engineers and not politicians or diplomats. A short answer may suffice. While engineers may not be directly responsible for the decision to go to war or decisions made in conducting a war, the ways in which war is conducted is due in no small part to their "ingenuity." Even decisions to go to war may be made based on considerations of military superiority due in large part to the work of engineers. Engineers can no more evade moral responsibility in virtue of being causally separated in varying degrees from the causes or conduct of war than can others who may be responsible for other types of conduct judged to be immoral. Neither can engineers—or anyone else—abrogate ethical responsibilities by pointing to others who may be more directly responsible. Like it or not, engineers (as do those in other professions and occupational fields) have duties that arise out of their specific roles as engineers.

Certainly not all who are in any way responsible will be *morally culpable* to the same degree. Yet there is good reason to believe that, whether for good or bad, engineers whose work is used for military purposes are morally implicated to a greater degree than, say, ordinary munitions factory workers or citizens whose taxes are used to support a war effort. Here, we may point to the chemical engineers responsible for the design, if not the actual manufacturing or deployment, of poison gas used in the Nazi death camps during World War II. The upshot is this. If a war is inherently unjust or cannot any longer be fought justly, then—just like anyone else—engineers who contribute directly to that war will, in varying degrees, be acting immorally. Because of the considerable overlap between engineering and military interests, engineers' casual connections to how wars are fought, and even if they are fought, are likely to be far greater than those of most non-engineers, excepting military planners and others employed directly in military decision making roles.

Importantly, this is not to suggest that such considerations rely on the view that war is always morally wrong or that engineering, even for military purposes, is intrinsically corrupt. We can imagine the case of those engineers who helped to fight the Nazis as a counterexample. Nevertheless, it is likely that, as a matter of circumstance, working as an engineer in certain contexts will involve one in practices that by one's own ethical standards are (or should be) regarded as immoral. Engineers neither choose the times they live in, nor do they generally determine the uses to which their work will be put. Yet as with all people in all professions, they retain an obligation to make moral judgments and to act ethically even when in circumstances and situations not of their own making.

Before we investigate the historical conditions underlying the contemporary context of most engineering work, and how individual engineers might best respond to that context, we will consider the ethics of war in general terms by reviewing just war theory.

2.2   JUST WAR THEORY

Just war—*bellum justum*—theory investigates the intersection of warfare and questions of justice.[7] Just war theory is usually divided into two distinct, though not wholly separable parts or questions. The first, *jus ad bellum* (justice of war), seeks to determine what conditions have to be met for a war itself to be just. In other words: When is it just to go to war? The second, *jus in bello* (justice in war), asks about the just conduct of war. Even if a particular war is just, that is, it meets the conditions of *jus ad bellum*, are there any moral constraints on how the war is to be conducted? If so, what are they?

The two aspects of just war theory are not unrelated. Absolute pacifists claim, generally on moral grounds, that there can be no such thing as a just war even in principle. However, there may be good reason, contrary to pacifism, to suppose that, while in principle the conditions for a just war (*jus ad bellum*) may be met, no actual war will in fact be just given that the conditions for *jus in bello* (justice in war), practically speaking, cannot be met. The idea here is that it is naïve to think that wars can be fought within the moral constraints required for justice in war. Such reasoning has become particularly relevant with the advent of nuclear weapons and other weapons of mass destruction, where for example, among other things, it is impossible to distinguish between combatants and non-combatants, which is seen as a condition that must be met for justice in war.

Even aside from indiscriminate weapons of mass destruction, it might be judged that other weaponry and technologies suppression and control (lethal and non-lethal), including those used for torture and surveillance, now make it impossible for there to be anything approaching justice in war. If this were the case, it would be due in no small part to the roles played engineers who designed such technologies.

2.2.1   *JUS AD BELLUM* – "JUSTICE OF WAR": WHEN IS IT JUST TO GO TO WAR?

Is it a mistake to talk about "just war" or justified war? Surely some wars are just even if they involve terrible acts—so long as a far greater good will come of it. Consider situations in which a war should have been fought—or fought earlier—but was not. Because recent wars have (arguably) been unjust insofar as they have not met the criteria for going to war justly in the first place, there may be an inclination to regard all war as unjust and immoral. Additionally, even if some recent wars were just, the atrocities carried out in such wars are arguably so great that they may be seen as violating not only conditions of justice in war (*jus in bello*), but also the conditions that have to be

---

[7] There are many fine studies of just war theory. See, for example, works by Johnson (1973; 1981; 1984; 1999), Walzer (1970; 1990; 2000; 2004), Phillips (1984: ix) for "a defense of the traditional position," Teichman (1986), Yoder (1984), and others in the references section for historical and contemporary perspectives on just war theory and for analyses of the conditions that must allegedly be met for *jus ad bellum* and *jus in bellum*.

met if going to war is to be just in the first place (*jus ad bellum*). Despite these grounds for assessing all war to be unjust, we make no such supposition here.

Phillips argues that the notion of just war (a morally justifiable war) only makes sense against a background of a moral point of view—one that has "moral prohibitions against [wonton] murder and acts of violence directed against others" (1984, 15–16). The question of "just war" would never arise if we did not think that something was ordinarily very wrong, morally speaking, with the kinds of actions that go on in war, such as killing. Assuming then that the kind of violence that occurs in war is generally wrong, what conditions have to be met for a war to be "just" (*jus ad bellum*), that is, for going to war in the first place?

There are variations of these conditions within just war theory, but the general conditions are as follows. For a war to be just it must be:

1. The *last resort*;

2. Declared by *legitimate authority* (ordinarily the state); and

3. *Morally justified*.

What constitutes "last resort"; who is a "legitimate authority"; and what constitutes the moral justification of war are all significant and important questions. There are no generally agreed upon answers to any of these questions.

Normative Ethics

Normative ethical theories try to tell us what is morally right to do and why. Such theories can be employed in an attempt to explain whether or not a particular war is morally justified and why. Different normative theories will always give different accounts of why some action is morally justified even when they agree, as they often do, that the action is justified. That is what makes them different theories. Three such normative ethnical theories are consequentialist, deontological, and virtue theories.[8]

- *Consequentialist theories* claim that the moral rightness or wrongness of an action is determined by the consequences—good or bad—of that action. J. S. Mill's Utilitarianism, for example, claims that an action is good if it produces the greatest good (defined as pleasure) for the greatest number.

- *Deontological theories*, such as those of Immanuel Kant, claim that an action is morally right no matter what its consequences, so long as it accords with some specified and inviolable ethical principles that the theory then goes on to elaborate.

[8] Brief accounts of normative ethical theories are often given in engineering ethics texts, as well as business ethics texts, and there are many excellent and accessible texts on ethics theory more generally. We believe it is more important, on practical as well as theoretical grounds, to be acquainted with ethics theory *as a guide to ethical reasoning* over, say, didactic ethical codes of conduct and practice.

- *Virtue theory* does not attempt to tell us, strictly speaking, what it is morally right, but focuses instead on the nature of good character and virtue. A virtuous person will tend to do the morally right thing, all things considered.

Assessing Justice of War

Determining if and when conditions of a just war are met is likely to be a contentious matter. For example, how can one be sure that going to war is, in the circumstances, really a last resort, and how is a war determined to be morally justified? Even the seemingly simpler second condition—that war must be declared by a *legitimate authority*—raises difficult questions. To what extent do states constitute legitimate authorities in declaring a war? Is it plausible to suppose that such authority universally resides only with states? And if so, how do we accommodate an oppressive state's citizens—who are certainly not recognized as legitimate authorities—rising up against their leaders?

The Nazis were democratically elected in the 1930s and may have been regarded by many observers as the legitimate authority of the German state at that time. Yet the war they declared was certainly unjust. Alternatively, it is implausible to suppose the war waged against the Nazis by civilian groups that were being slaughtered was unjust, even as the status of those groups as a legitimate authority is unclear. As Phillips suggests, to accept the condition that war must be "declared by a legitimate authority," and no other party within the jurisdiction of that authority, as a necessary condition of just war, one must assume that such an authority will "protect those for whom it has a political responsibility" (1984, 16).

The condition that a just war must be *morally justified*, while virtually true by definition (what else does "just" mean?), is likely the most problematic of all. The two broadest and most commonly identified moral justifications for war are *self-defense* and *efforts to avoid greater injustices*. But here, again, difficult questions arise as to the conditions constituting self-defense. For instance, under conditions of civil strife, at what point does majority-rule become minority oppression justifying self-defense measures? At what point do political impasses demand military response as a matter of self-defense? Even more open-ended is the question of when going to war "to avoid greater injustices" is justified. This is virtually always a matter of dispute—and nowhere more than in recent history. Just think of the French and then U.S. warfare in Vietnam or the Gulf and Iraqi wars or the justification of the Syrian regime's attacks on its own citizens—as terrorists, the late modern incarnation of "evil"—in 2012.

Just war theory has evolved, leading Phillips to distinguish between traditional and modern accounts. Summing up their different positions on what constitutes justice of war, Phillips argues:

> Wars of aggression *are permitted under traditional doctrine only if the cause is just; but all wars of aggression are prohibited under the modern interpretation, for no matter how serious the injury to a state, modern warfare is an immoral means for settling grievances and altering*

existing conditions.… [T]he destructiveness of modern warfare makes it a wholly disproportionate means for the resolution of international disputes...even where they are just.… [Additionally], to admit the right of states to initiate combats, even to correct injustice, would impede efforts of the world community to establish a judicial method of outlawing war.… A war of defense against the injustice of aggression is morally permissible in both the traditional and the modern view. This is perceived as in no way a contradiction of the concern for peace, for peace may require defense. (1984, 26–7, emphasis added)

Phillips also says "the traditional doctrine of *bellum justum* positively requires that states extend military help to victims of injustice" (1984, 22). This raises additional questions, including when, under what circumstances, and to what extent others are morally required to intervene—or short of being "required to intervene" nevertheless have a right to intervene. Phillips's view is that "there is no general formula...but...[we] may say that a state has the right to intervene...where there is an overt and systematic program of injustice… [though] a state need not, however, have a positive duty [e.g., be morally required] to do so" (1984, 22).

Engineers' Moral Responsibilities

Even if engineers reject the criteria set out above and substitute other reasoned grounds, they, like everyone else who may play a role, have an obligation to judge whether a war is ethical. The fact that an engineer working on a nuclear weapon may have more of an obligation, and a more immediate obligation, to think about the possible applications of his work, some of which may be difficult or impossible to foresee, than an engineer involved in the design of tires, does not mean that the latter has no responsibility regarding what he works on and how it may be used.

It is entirely possible that, for all one knows, tires or some other apparently innocuous product that engineers have a role in producing, rather than nuclear weapons, may have the more significant role to play in particular conflicts. This raises an important question (one that will be returned to in later chapters): Given that an engineer has little control, and rarely exclusive control, over how the things she helps design and produce are used, to what extent can she be held morally responsible for their use or what results from that use? Answering this question requires attending to causation-at-a-distance.

The moral responsibility surrounding the idea of *causation-at-a-distance* may seem difficult to accept or even understand, and yet it is vitally important. It may be the case that the very possibility of any particular war is usually distant from most of the relevant engineering decisions that provide the overall capability of going to war. Except for engineers working directly on military projects, it may be very hard to know if one's work has military potential, not to mention to anticipate the conditions that will lead to its application in a military confrontation. Nevertheless, the important point is that in engineering (as elsewhere) being somewhat removed (temporally and spatially) from more immediate and direct causes of immoral actions and their effects does not necessarily absolve one from all moral responsibility and culpability for what occurs.

Indeed, responsibility and moral culpability can outrun intention and, even, what one knows. Persons may be responsible for an action or result that they do not intend or even reasonably foresee on the basis of what they currently know. We can see a person as "culpably ignorant" about some fact or possibility if that person might reasonably be expected to know those things. After all, if intention or knowledge were necessary in order to be held morally accountable for some outcome, then one could avoid responsibility by willfully keeping oneself in the dark about relevant issues or refusing to entertain certain possibilities.

What we reasonably expect a person to know or to "look into" is partly a function of how important the matter is and what the consequences might be. It is also partly a function of organizational and disciplinary division of labor, where we expect due process and relevant expertise employed in cases of importance and consequence. We reasonably expect civil engineers to know if they are building a bridge in an area particularly prone to earthquakes and if so to build it accordingly. Knowledge of geologic seismic activity is taken for granted among engineers. Extending the metaphor, engineers should also have knowledge of social, economic, and political faultiness.

While the precise conditions of morally culpable ignorance may be difficult to specify, moral innocence cannot be maintained merely by refusing to come to know what one should know or to think about what, in the circumstances, one should think about. Self-deception—denying or avoiding reachable relevant knowledge—is a common cause of culpable ignorance. Among other things, and particularly when feeling morally conflicted, we should be aware of the values we allegedly adhere to, as well as the pressures we face to act in ways contrary to those values.[9]

The Case of Nuclear Weapons Development

As a case in point, given what an engineer might reasonably be expected to know and think about regarding the technology's possible use or misuse, the nuclear engineer or scientist who works on nuclear weapons cannot plausibly claim that they have *no* responsibility (whether praiseworthy or blameworthy) for how those weapons are used (whether detonated or for diplomacy) and the results. While the nature and extent of their responsibility is open to debate, the fact that such responsibility exists is not. Perhaps the most dramatic case surrounding this technology, one where moral responsibilities were discussed and embedded in related issues of security, patriotism, duty, human nature, human welfare, and the like, was that of J. Robert Oppenheimer (1904–1967), the so-called "father of the atomic bomb."

---

[9] See McCuen and Gilroy (2010, 67–103), Chapter 4, "Value Decision Making." See also Catalano (2006, 20) for an account of engineering as "a value-laden profession" that frequently fails to recognize, let alone seriously act upon, values, commitments, and principles.

On July 16, 1945, in New Mexico, the first atomic bomb was detonated. After the test, code-named "Trinity," Oppenheimer reported he thought about the following quotation from the Bhagavad Gita: "Now, I am become Death, the destroyer of worlds."[10] Many years later he said, "I had never said that I had regretted participating in a responsible way in the making of the bomb" (Seagrave, 1964). One wonders whether Oppenheimer's claim not to regret making the bomb extends to its deployment—not regretting, or not claiming any responsibility for, the bomb being dropped on the (non-combatant) residents of Hiroshima and Nagasaki. But regardless of his reflections, he was involved in the causal chain that resulted from Truman's decision to drop the bomb, a decision that undoubtedly shortened the war, but did not alter its ultimate outcome.[11]

To what extent was Oppenheimer responsible for the bomb being dropped? Given that he was at least partly responsible for "inventing" it, there are the further questions of whether the part he played was morally blameworthy or praiseworthy and whether he should have regretted his role in the production of the bomb, as well as the production of the bomb generally in the circumstances at the time, even if he did not.[12] Undoubtedly, given the larger historical and institutional context, if Oppenheimer had not played the role he did, then someone else would have been enrolled to play a similar role—either then or some time later. But this fact does nothing to mitigate Oppenheimer's own responsibility—no matter what we judge that responsibility to be.

The fact that relatively few engineers are involved so directly in matters of such enormous consequence (though it should be remembered that a significant absolute number are) does nothing to change the question of moral responsibility on the part of scientists and engineers for their actions—direct and indirect, individually and as a group (Naverson, 2002; Gilbert, 1997). The fact that generally many others are involved in such projects as well does not alter the moral issues.

With regard to less controversial technologies, say, automobiles and planes, responsibility for foreseen deaths by accidents is generally mitigated by factors and persons more directly casually linked to any given accident (e.g., weather, careless driving). In other cases, such as the infamous faulty gas tank on the early 1970s Ford Pinto or tread separation problems with early 2000s Firestone tires, the responsibility for ensuing injuries and deaths is again laid at the feet of those who, in these cases knowingly, decided upon courses of action that resulted in such accidents. Moral culpability is especially obvious for those who refused to have the cars and tires taken off the market after the problems became apparent, but moral culpability also extends to those who did not speak up about the faults and dangers, whether due to monetary or other job-related considerations.

---

[10] A video transcript of Oppenheimer's reflections is available at The Atomic Archive (accessed May 21, 2012): http://www.atomicarchive.com/Movies/Movie8.shtml.

[11] Curiously, more recent research suggests that Oppenheimer's original response to the Trinity detonation was jubilation—surprise and triumph that it had actually worked—rather than somber reflection (Monk, 2012).

[12] See Van de Poel and Royakkers (2011), Chapter 9 (pp. 249–276), "The Distribution of Responsibility in Engineering" and Chapter 8 (pp. 217–248), "Ethical Aspects of Technical Risks."

Engineering can play an important, albeit nonobvious, role in determining whether it is just to go to war in a particular case (*jus ad bellum*). In general terms, engineering expertise is frequently relied upon to make a political and moral case for war as last resort and war as morally justified. For example, engineers (and scientists) help determine the nuclear capability of a state or group or the presence of weapons of mass destruction, and they report their assessments to political and military leaders. Similarly, these leaders make strategic and tactical decisions in pursuing war based on the technological capabilities of their own militaries as assessed in collaboration with engineers. Such information may be correct or incorrect and may be used rightly or wrongly (wrongly, in the case of the supposed weapons of mass destruction justifying the 2003 invasion and occupation of Iraq) to justify pre-emptive strikes on the ground of "last resort." At the same time such information may serve as part of a moral justification for the war itself on the ground of "self defense."

2.2.2   *JUS IN BELLO*— "JUSTICE IN WAR": HOW CAN WAR BE FOUGHT JUSTLY?

We turn now from the question of when it is just to go to war (*jus ad bellum*) to the question of the just conduct of war (*jus in bello*). It is here that engineering can be shown to be particularly relevant. What, if anything, constitutes *justice in war*? The focus will again be the relevance of this issue to the ethics of engineering practice. The conditions for justice in war, or conducting a war justly, include the following two criteria: proportionality with regard to the force used in war and discrimination between combatants and non-combatants.

Proportionality

The principle of *proportionality* requires that the force employed or threatened must be morally proportionate to the end sought. For the use of force to be justified, it must be proportional to the injustices that the use of force seeks to rectify, and not create or result in even greater injustices. (This requirement is also relevant in determining whether a war is morally justified in the first place.) For if the use of force must be excessive, or one knows it will be excessive—think of all-out nuclear retaliation—then, arguably, the war cannot be justified and neither can the use of such force.

The existence of nuclear weaponry, and in particular the destructive capacity of nuclear arsenals amassed during the Cold War, poses serious challenges to the prospects of conducting war justly under the principle of proportionality. However, the idea that nuclear weapons have made war "unthinkable" in ways that render just war theory obsolete seems to have faded with the end of the Cold War. Even post-September 11, 2001, and with the advent of "terrorist" threats combined with the nature of the governmental responses to those threats, populations at large are not today preoccupied with the possibility of nuclear obliteration in the ways they were at the height of the Cold War in the 1980s. This is so despite nuclear arsenals—including "tactical" nuclear weapons, modes of delivery, etc.—being far more powerful and technologically advanced than ever before. Nevertheless, just war theory does bring these possibilities to the fore as something that must be considered if the role of engineers in relation to war, and the ethical implications of contemporary war for engineering practice, is to be understood and acted upon.

The idea that the force used in war must be proportional and not excessive is not the same as the idea that the force used must be the minimal force necessary to achieve one's objective—whether in a battle or with the war itself. This is because the minimal force necessary to win a battle or war (nuclear obliteration of an opponent—including non-combatants) may well exceed the force that the principle of proportionality would allow. Phillips says, "If one knows that certain sorts of weapons will be used which one knows will cause casualties that are disproportionate on any objective basis, it is obviously morally preferable to attempt to obtain whatever proportionality is possible relative to the system—even in the case of massive retaliation with nuclear weapons" (1984, 29).

But it is not at all clear that acting in such a way is "obviously morally preferable" given that, with respect to, say, all-out nuclear war, no proportionality whatsoever is possible relative to the system. Few suppose that killing in a justifiable war is never justifiable. Yet, given the conditions of modern warfare, many do believe no war can be justified. Their claim is that nuclear warfare, or even technologically advanced non-nuclear warfare, renders proportionality unattainable. The same holds true for the principle of discrimination.

Discrimination

The *principle of discrimination* holds that force must never be used in a way that makes either non-combatants or innocent persons the intentional objects of attack. According to this principle, the only morally legitimate targets in war are combatants. If the application of the principle of proportionality is difficult, then that of the principle of discrimination is even more so, incorporating the most intractable challenges of proportionality and adding others. Determining who is and who is not a combatant poses a considerable challenge, and this challenge has become more complex in recent years with the shifting of warfare toward economic, psychological, and technological domains.

With these changes, it is less plausible to define combatants only as soldiers who are actually committing violence (i.e., fighting). (And if there is a problem distinguishing between combatants and non-combatants in theory, think of how much more difficult it is in practice.) Adding to the complexity, the combatant/non-combatant distinction is not the same as the distinction between those who are morally responsible for the war and those who are not. Combatants may have less moral responsibility for their actions than many non-combatants. The principle of discrimination explicitly recognizes that not all non-combatants are innocent and that not all combatants are "guilty" or morally culpable (i.e., some may be "innocent").

Supposing a way can be found to reasonably distinguish, at least in theory, between combatants and non-combatants, one is still faced with the problem of targeting the former without targeting the latter. Cluster bombs and land mines, by their very nature, are likely to target non-combatants. Yet the principle of discrimination requires that force must never be used in a way that makes non-combatants the intentional objects of attack. In many instances of war, it is not practically possible to target combatants without also harming non-combatants, even if unintentionally. This may be seen to be a particular problem in modern warfare, where armies no longer face each other on discreet battlefields but instead fight in urban centers and other places where targeting of only combatants may be impossible.

Assessing the extent of justice in war requires a series of interpretations to be made, each of which is fraught with difficulty. First, one must determine what the conditions of proportionality and discrimination entail in a given case of war. Then, one must decide whether the two conditions have been met or violated, or—given the nature of modern warfare considered both politically and technologically—whether the two conditions even *can* be met. According to the justice-in-war principle, if these conditions cannot both be met—that is, if a war cannot be fought in a just manner—then even if the justice-of-war principle has been met, the war itself cannot be just. If a war cannot be fought justly, then the justice of the war is irrelevant. Both conditions of justice in war—proportionality and discrimination—are applicable and relevant when considering a war's morality and, by extension, the moral responsibilities of engineers whose work contributes to such wars.

Focusing on the requirements of proportionality and discrimination provide another lens with which to view the ethics of engineering practice surrounding warfare. If, as some claim, neither of these conditions can be met in modern warfare, this is in no small way due to the nature and capabilities of weapons, systems, and countless ancillary products and measures, produced in part by engineers, that shape and structure the form modern warfare has assumed.

2.2.3   JUST WAR AND ENGINEERING ETHICS

Aside from the obvious case of the design and production of weapons and weapons systems, where many fields of engineering play a role, engineering is linked to warfare in other ways. In fact, without engineering, there could not be war as we know it today. This is not meant to imply that engineers are fundamentally or primarily *responsible for war*. But it does suggest that engineering is heavily and necessarily implicated in the shape, structure, procedures, and conduct of war. As with the case of justice of war, the fact that others are responsible in various ways, to greater or lesser degrees, is irrelevant morally speaking. That others may be responsible for making it impossible to meet the conditions of justice in war, or that others would surely take the place of oneself in the casual chain, in no way mitigates one's individual moral responsibilities.

Importantly, moral responsibility encompasses both moral praiseworthiness as well as moral culpability. So to say that engineers (or engineering) are "responsible" for war in the ways mentioned is not necessarily to suggest that such responsibility is morally blameworthy. Engineers who play roles in wars that are just and necessary in the circumstances (assuming there are any) may be judged to be acting in a morally praiseworthy manner. The overarching point, then, is that engineering and warfare are intrinsically linked, whether or not a particular war is just or the actions of engineers in the service of a war are morally right or wrong.

The nature of the specific weaponry and warfare systems that engineers are directly involved in producing may render the possibility of meeting the requirements of justice-in-war, proportionality and discrimination, as well as justice-of-war a thing of the past. So called "smart-bombs," laser guided missiles, and other high-tech weaponry are meant to address the problem of indiscrimination—otherwise known here by the misnomer of "collateral damage"—and they are sometimes successful to a degree in doing so. But the fact remains that modern warfare makes it extremely difficulty to distinguish combatants from non-combatants and innocent non-combatants from those who willingly and directly participate in warfare.

Given this condition, just war theory suggests, engineers and others should distance themselves from direct and indirect participation in war. Supposing, on the other hand, that the possibility of a just war remains, engineers still need to consider their individual ethical responsibilities in regard to the war. They must ask how they can respond with integrity to professional work contexts oriented toward warfare.

CHAPTER 3

Engineering Integrity

Just war theory can provide conceptual tools to identify key questions surrounding engineering contributions to war in broad terms, highlighting the need to be aware of long and indirect chains of causation. An elaborated *theory of integrity* can also help engineers (and others) navigate their professional responsibilities, and respond appropriately to militarism and warfare as it intersects with their lives and their decisions in complicated ways, and as each of us tries to make ethically sound, self-aware decisions.

This chapter is intended to address the challenges faced by individual engineers in confronting the complex implications of their lives and decisions, constrained as they are by structures and circumstances well beyond any individual's control. It carefully explores what integrity means for individuals confronting life-course decisions and then explores how the concept of integrity can be useful for navigating the conceptual and institutional structures that delimit engineering practices.

3.1 WHAT IS INTEGRITY?

To open up our analysis of integrity, we will look at the hypothetical case of Georgina.[13] Georgina recently received her Ph.D. in chemical engineering and is having difficulty finding work. Her health is poor which adds to the difficulty in finding a job. Georgina's spouse works to support their young and struggling family, which includes two young children. She is something of a pacifist and is certainly opposed to chemical and biological warfare. She is told about a well-paying job in a laboratory working on biological and chemical warfare. She has no illusions about her indispensability and knows that, if she does not take the job, then someone else will. In all likelihood, it will go to someone without pacifist commitments, who will pursue the research without her ethical qualms.

There are any number of combinations and permutations of this scenario—some morally relevant in the situation and others not. We can imagine "George" as a man, rather than as woman. We can imagine she already has a job but a big promotion is dependent upon doing research that Georgina loathes. Or we can imagine things are even harder: That she is a single parent or that she lives in a dictatorship with high unemployment and rampant poverty.

Although it could be argued both ways, the typical utilitarian view of her situation is that Georgina should take the job, because it results in a greater overall good, regardless of its impact

[13] This example is adapted with changes from Bernard Williams (1973, 97–99). Parts of this section also draw from Cox et al. (2003). Williams's example used the man's name George, but in the interest of flagging engineering's traditional masculinism, we have elected to use Georgina in our example. Suffice to say that many ethical issues will face women and men alike, women and men will face them differently, given the ways that gender difference affects social interaction and obligation.

on her character. But if we ask the question, "Should Georgina take the job?," not as a question of normative ethics, but also in relation to *her own perspective* (that is, in the light of her own beliefs, commitments, knowledge, value, and desires), then we arrive at a different answer. Many of the features of her situation are likely to weigh in, and it is appropriate that Georgina consider these features to the extent that she/he is able. When we ask whether Georgina should take the job as she would ask it—"Should I take this job?"—we must consider the extent to which taking the job would undermine her own sense of what she should do. Would it undermine her integrity to take the job? Would she think less of herself upon taking it? Would those who know Georgina think less of her if she took it? These are questions of Georgina's integrity.

While integrity is regarded as an important virtue, its meaning is difficult to pin down. Most generally, "integrity" is understood as acting consistently with one's values. It is at times used synonymously with being "moral," but acting morally is not same as acting with integrity, since one may acknowledge a person is acting with integrity even though that person holds importantly mistaken moral views. Hence, a person's action may be consistent with that person's values, even though those values and, therefore, the action are immoral. In the case of Georgina, we assume that what is troubling about a decision to take the job is that Georgina feels she would be abandoning important commitments, and would not be "true to herself" doing so. Part of the reason for this is that Georgina knows the job would require her to act immorally, or at least contrary to her beliefs around non-violence.

Simply rejecting the job, however, is not easy, because Georgina experiences other, competing strong desires, commitments, beliefs, and values. If these were not also at play, she would have no problem rejecting the weapons job offer. Because of these competing desires, commitments, beliefs, and values, she may delude or deceive herself in a variety of ways that serve to resolve the tension, but also undermine Georgina's integrity. For example, Georgina might "talk herself into" taking (or not taking) the job by manufacturing wish-fulfilling but unwarranted beliefs about the nature of the job or by seeking out excuses that are plausible but not derived from her own sense of identity—for example, using the utilitarian logic: "If I don't take it, someone else will." How then does Georgina, conflicted as she is, decide the issue in a way that maintains her integrity?

To understand the situation Georgina faces, recourse to codes of ethics or simplistic moralizing will not suffice. Georgina's problem has to do with what is being asked of her—in terms of moral, professional, and technical practice—and how that might conflict with her self-conception: who she is and thinks she should be; what she stands for; and, in a sense, what she lives for. On this view, integrity is an excellence and consistency of character. It is also a way of dealing with self-conflict, temptation, ambivalence, the competing demands of relationships and social roles, and the need for one's commitments to sometimes change over time in view of new desires, knowledge—including self-knowledge—beliefs, values, and the like.

Integrity is often attributed to various aspects of a person's life; for example, we speak about professional, intellectual, and artistic integrity. Ordinary discourse about integrity involves two fundamental intuitions: first, that integrity is primarily a formal relation one has to oneself; and second, that integrity is connected in an important way to acting morally. In other words, there are some substantive or normative constraints on what it is to act with integrity.

However, the most important sense of the term relates to general *character*, and in defining integrity of character, philosophers have tended to emphasize one of the above intuitions at the expense of the other. They have advanced five major approaches to assessing the integrity of a person's character: integrity as self-integration; integrity as maintenance of identity; integrity as standing for something; integrity as moral purpose; and integrity as a virtue. In order to grapple fully with the range of variables underlying an assessment of Georgina's career decision, we review each of these approaches to understanding integrity.

3.1.1 INTEGRITY AS SELF-INTEGRATION

The self-integration view of integrity sees integrity as a matter of persons integrating various parts of their personality into a harmonious whole. Understood in this way, the integrity of persons is analogous to the integrity of things: integrity is primarily a matter of keeping the self intact and uncorrupted. In attributing self-integration to a person, we are making no evaluative judgment of the states that are integrated within the person. Whether one's intentions are worthwhile, or whether one's judgment or reasoning is sound, makes no difference on the "self-integration" account. On this account, so long as a person has views, beliefs, desires, and values that accord with one another, and there are no internal constraints (conflicts) to being able to willfully act accordingly, that person has integrity.

According to Harry Frankfurt (1987, 33–34), desires and volitions (acts of will) are arranged in a hierarchy. First-order desires are desires for various goods; second-order desires are desires that one desire certain goods, or that one act on one first-order desire rather than another, and so on. According to Frankfurt, wholly integrated persons bring these various levels of volition and desire into harmony and fully identify with them. Such identification appears to involve knowing them, not deceiving oneself about them, and (usually) acting in accord with them. A person, as Georgina's situation highlights, is subject to many conflicting desires in career and life. If one acted at each moment according to the strongest current desire, with no consideration of or discrimination among more or less important desires, then one would inevitably act without integrity.

Thus, integrity requires one to discriminate among different-order desires, endorsing some at the expense of others, usually by identifying those of a higher order, for example, by ranking long-term success over short-term fun. Since desires inevitably conflict, people require strategies for negotiating such conflicts. Full integration of one's self requires one to decide upon a certain structure of desires—arranged as a hierarchy, so one's lower-order desires and volitions do not confound achievement of high-order ones. As Frankfurt puts it, when a person unreservedly decides to endorse a particular desire,

> ...the person no longer holds himself [sic] at all apart from the desire to which he has committed himself. It is no longer unsettled...whether the object of that desire...is what he really wants: The decision determines what the person really wants by making the desires upon which he decides fully his own. To this extent the person, in making a decision by which he identifies with a desire, 'constitutes himself.' (1987, 38)

When agents thus constitute themselves without ambivalence or inconsistency (that is, unresolved desires for incompatible things), they have what Frankfurt calls "wholeheartedness."

On one way of developing this view, wholeheartedness is equated with integrity. It should be noted, here, that self-conflict is not limited to conflict among desires. Conflicts also exist among different-order commitments, principles, values, and wishes. Furthermore, all of these things—desires, commitments, values, and so on—are in flux. They change over time, so achieving the kind of wholeheartedness Frankfurt describes is a never-ending process. Self-knowledge is crucial to this process insofar as, for example, one must know what one's values are so one can first order them and then act appropriately.

A number of criticisms have been leveled against the integrated-self view of integrity. First, it places only formal limits on determining *what kind of person* has integrity, and does not apply substantive constraints, such as honesty and genuineness. On Frankfurt's account, a person may be perfectly integrated, be "wholehearted," and thus have integrity, while being a cheater or liar, so long as the person cheats or lies with conviction and consistency.

Second, the integrated-self view of integrity places only formal limits on *the kind of desires* that constitute a self, and again does not impose substantive constraints. So, in Georgina's case, a desire to impress friends with the title of her position at the laboratory could reasonably be more important, on Frankfurt's account, than a desire to support her family or to enact her commitment to pacifism.

Third, on some accounts of the integrated-self view, the fully integrated person would never experience genuine temptation. Temptation requires that the full force of an "outlaw" desire be experienced, but successful integration of the self may mean that such desires are fully subordinated, precluding a person experiencing them. That a person experiences, and overcomes, temptation would count against their integrity on such a view. Common sense suggests that a capacity to overcome temptation, and thus to enact strength of character, is an essential element of integrity, not its lack (Halfon 1989, 44–7).

Fourth, as Calhoun (1995, 238-41) argues, people can find themselves in situations where wholeheartedness tends to undermine rather than constitute their integrity. For example, wholeheartedness may preclude one from further considering options, values, perspectives, information, and other things that should be considered in the circumstances. Analogously, Davion (1991, 180–

192) argues that a person may change radically—change even some of their fundamental values and desires, for example—and yet maintain integrity. In the midst of what are inevitably complex and multifaceted lives, people may have compelling reasons to avoid neatly resolving incompatible desires. The cost of such resolution may be a withdrawal from those aspects of life that make genuine claims upon us. Resolving self-conflict at the expense of deep engaging with all the different facets of one's life does not seem to contribute to one's integrity. One should not confuse integrity with mere decisiveness or neatness in responding to complex circumstances.

3.1.2   INTEGRITY AS MAINTENANCE OF IDENTITY

Related to the integrated-self view of integrity is an approach that understands integrity to be "holding true to one's commitments." On this view, a "commitment" covers many different kinds of intentions, promises, convictions, and relationships of trust and expectation. We are often committed in different ways to many kinds of things: people, institutions, traditions, causes, ideals, principles, projects, and so on. Commitments can be explicitly, self-consciously, publicly entered into or implicit, unselfconscious, and private. Some are relatively superficial, while others are very deep.

As with the integrated-self view of integrity, however, this approach emphasizes formal conditions over substantive content, here the content of one's commitments. Given that people have so many commitments, of many different kinds, and because commitments clash and change over time, defining integrity merely in terms of remaining steadfast to one's commitments is dubious. It matters *which* commitments one remains true to. Philosophers have developed different accounts of integrity in response to this need by specifying which kinds of commitments should be centrally important in assessing a person's integrity.

One such account is to define integrity in terms of the commitments that people identify with most deeply, the commitments that align with what they consider their lives to be fundamentally about. Commitments of this kind are sometimes called "identity-conferring commitments" or "ground projects." This view of integrity, the identity view, is associated most closely with Bernard Williams (1973; 1981b). The idea is that abandoning an identity-conferring commitment is to lose grip on what gives one's life identity or individual character. An identity-conferring commitment, according to Williams, is "the condition of my existence, in the sense that unless I am propelled forward by the conatus of desire, project and interest, it is unclear why I should go on at all" (1981b, 12).

Hence, on the identity view of integrity, to act with integrity is just to act in a way that accurately reflects your sense of who you are—to act from motives, interests, and commitments that are most deeply your own (Williams, 1981a, 49). However, if integrity is understood to be the maintenance of identity-conferring commitments, then integrity cannot really be a virtue. This is Williams's view. He argues that integrity is not related to motivation as virtues are. A virtue either motivates or enables a person to act in desirable ways. But if integrity is no more than maintenance of identity, then it can play neither of these roles. Additionally, on this view of integrity, there appears to be no normative constraints, either on what such commitments may be or on what the person of integrity can do in the pursuit of these commitments. People of integrity could do horrific things and maintain their integrity so long as they are acting in accordance with their core commitments.

There are additional problems with the identity view of integrity. Integrity is usually regarded as something worth striving for, which the identity account of integrity fails to make sense of. It disconnects integrity from virtue and, hence, from being praiseworthy. Additionally, the identity theory of integrity ties integrity to commitments with which a person identifies, but acts of identification can be ill-informed, superficial, or foolish. People may, through ignorance or self-deception, fail to understand or properly acknowledge the source of their deepest commitments and convictions. On the other hand, this view of integrity as maintenance of identify-conferring commitments does recognize the relevance of self-knowledge to acting with integrity. If people fail to act on their core commitments, through self-deception, weakness of will, cowardice, or even ignorance, then to this extent they may be said to lack integrity.

Further, on the identity view of integrity, a person's integrity is only at issue when their deepest, most characteristic, or core convictions are brought into play. This contradicts the expectation that persons of integrity behave with integrity in a variety of contexts, not only those of central importance to them (Calhoun, 1995, 245). Like the self-integration view, the identity view of integrity places only formal conditions upon the kind of person that might be said to possess integrity, and so a similar criticism can be made of it on this ground: Integrity demands placing certain limits on the *kinds of commitments* a person holds.

3.1.3   INTEGRITY AS STANDING FOR SOMETHING

Contrary to the self-integration and identity views of integrity, which see it as largely a personal virtue defined by a person's care of the self, Calhoun (1995) argues that integrity is primarily a *social virtue*, defined by a person's relations to others. Persons of integrity do not just act consistently with their own endorsements; they stand up for their best judgment within a community of people trying to discover together what in life is worth doing. As Calhoun says,

> Persons of integrity treat their own endorsements as ones that matter, or ought to matter, to fellow deliberators. Absent a special sort of story, lying about one's views, concealing them, recanting them under pressure, selling them out for rewards or to avoid penalties, and pandering to what one regards as the bad views of others, all indicate a failure to regard one's own judgment as one that should matter to others. (1995, 258)

On this view, integrity entails not only that one stand up, without hypocrisy, for one's best judgment, but also that one has proper respect for the judgment of others.

Calhoun's account seeks to avoid the problem of distinguishing between fanatical zeal and integrity. Fanatics are able to integrate desires, volitions, and beliefs to their own satisfaction in, what seems to them at any rate, a coherent manner. They remain steadfastly true to their deepest commitments like no others. Whereas the self-integration and identity views of integrity threaten to make the fanatic a paradigm case of integrity, directing attention to social judgment countervails this tendency by requiring respect for the deliberations of others.

But Calhoun's concept of "proper respect for others' views" raises other questions. What amounts to "proper" respect? And which others' views? Calhoun clearly does not mean always agreeing with some larger social group. Exemplary figures of integrity often stand by their judgment in the face of enormous social pressure. How, then, is one to differentiate between fanaticism and standing up for one's views under the pressure of others? Calhoun's claim that fanatics lack integrity because they fail to properly respect the social character of judgment and deliberation sounds reasonable, but that is because most of the work is done by the idea of "proper respect"—and it is not clear in the end what this comes to.

Like the other views thus far considered, Calhoun's account of integrity places no substantive constraints on the kinds of commitments that a person of integrity may endorse. It does not seem necessary on her view that a person of integrity has a special concern with acting morally. Although they have a special concern to understand what in life is worth doing, persons of integrity are not constrained to give moral answers to this question that take others' thoughts into account. By contrast, the following account of integrity is explicitly concerned with attitudes toward morality.

3.1.4 INTEGRITY AS MORAL PURPOSE

Another way of thinking about integrity places moral constraints upon the kinds of commitment to which a person of integrity must remain true. There are several ways of doing this. Elizabeth Ashford (2000, 46) argues for a virtue she calls "objective integrity." Objective integrity requires that agents have a sure grasp of their moral obligations. A person of integrity cannot, therefore, be morally mistaken. Understood in this way, one can only properly ascribe integrity to a person with whom one finds oneself completely in moral agreement. This concept of integrity does not, however, match ordinary use of the term, which entails honoring the principles upon which someone stands even if one disagrees with those principles. In this usage, the point of attributing integrity to another is not to signal moral agreement, but instead to open a space for substantial moral disagreement without launching a wholesale attack upon the other's moral character.

Mark Halfon offers a different way of defining integrity in terms of moral purpose. Halfon (1989) describes integrity in terms of a person's dedication to the pursuit of a moral life and their intellectual responsibility in seeking to understand the demands of such a life. Halfon says that persons of integrity:

embrace a moral point of view that urges them to be conceptually clear, logically consistent, apprised of relevant empirical evidence, and careful about acknowledging as well as weighing relevant moral considerations. Persons of integrity impose these restrictions on themselves since they are concerned, not simply with taking any moral position, but with pursuing a commitment to do what is best. (1989, 37)

This view allows an approach to integrity that is not necessarily "objective," as Ashford's approach seeks to be. Like Calhoun, Halfon sees integrity as centrally concerned with deliberation about how to live. Halfon, however, conceives this task in narrowly moral terms and ties integrity to intellectual virtues exercised in pursuit of a morally good life. He speaks of a person confronting "all relevant moral considerations," but what counts as a relevant moral consideration, on Halfon's view, depends upon the moral point of view of the individual moral agent. Persons of integrity may thus carry out acts others would regard as grossly immoral. What is important is that they act with moral purpose and display intellectual integrity in moral deliberation.

Halfon (1989: 134–6) admits that this approach to integrity makes possible, however unlikely, a Nazi bent on genocide to be a person of moral integrity. Other philosophers object to this consequence. If it is anyhow possible for the genocidal Nazi to be ascribed with moral integrity, they argue, then we could likewise ascribe integrity to people whose moral viewpoints are bizarrely remote from any we find intelligible or defensible (see McFall, 1987; Cox et al., 2003, 56–68).

Daniel Putnam (1996) draws on the work of Carol Gilligan (1982) to suggest a different way of addressing the problem of the Nazi of integrity. Moral constraints upon attributions of integrity need not take the form of Ashford's "moralized" view or Halfon's more limited formal view. Instead, one can hold that attributions of integrity involve the judgment that a person acts from a moral point of view those attributing integrity find intelligible and defensible (though not necessarily right), and that this constraint does have substantive implications. It prohibits attributing integrity to, for example, those who advocate genocide or deny the moral standing of people on, for example, sex-based or racial grounds.

We accept the line of reasoning that there are things a person of integrity cannot do. The Nazis and other perpetrators of tremendous and deliberate harm were either committed to what they were doing, in which case they were profoundly immoral and lacked integrity, or else they lacked integrity because they were self-deceived or dissembling and never actually had the commitments they claimed to have. Judgments of integrity would thus involve judgment about the reasonableness of others' moral points of view, rather than the absolute correctness of their view (as per Ashford) or the intellectual responsibility with which they generally approach the task of thinking about moral questions (as per Halfon).[14]

Defining overall integrity of character in terms of moral purpose has the advantage of capturing intuitions of the moral seriousness of questions of integrity. However, the approach appears

---

[14] See McFall (1987) for a discussion of the nature of the constraints on proper attributions of integrity.

too narrow. Halfon's identification of integrity with moral integrity appears to leave out important personal aspects of integrity, aspects better captured by the other views of integrity we have examined. Integrity does not seem to be exclusively a matter of how people approach evidently moral concerns. Other matters like love, friendship, and personal projects appear highly relevant to judgments of integrity. When we judge Georgina's integrity overall—as a quality of excellence of her character—or she judges herself, the decision is not likely to be based solely on whether she has moral integrity.

3.1.5 INTEGRITY AS A VIRTUE

The accounts of integrity we have examined thus far have a certain intuitive appeal and capture important features of the concept of integrity. There is, however, no philosophical consensus on the best account. It may be that the concept of integrity is a cluster concept, tying together different overlapping qualities of character under the one term. Taking this approach, Cox et al. (2003) argue that integrity is a virtue, but not one that is reducible to the workings of a single moral capacity (in the way that, say, courage is) or to the wholehearted pursuit of an identifiable moral end (in the way that, say, benevolence is).

One gains a fair grasp of the variety of ways in which people use the term "integrity" by examining conditions commonly accepted to defeat or diminish a person's integrity. Integrity—like other virtues according to Aristotle—stands as a midway point (a mean) between two excesses—one maintaining imperfect conditions through inaction and the other through acting but doing so in a way that undermines the stability needed for productive change. In this sense, "courage" is midway between timidity or undue concern as one excess and foolhardy recklessness as an opposing excess. For clarity, we elaborate each of these excesses.

On the side of reluctance to enact change are things like arrogance, dogmatism, fanaticism, monomania, preciousness, sanctimoniousness, and rigidity. These are all traits that can defeat integrity insofar as they undermine and suppress attempts by individuals to critically assess and balance their desires, commitments, wishes, changing goals, and other factors.

On the other side, a different set of characteristics undermine integrity. These do not undermine the status quo as much as they make it impossible to discern stable features in one's life, including one's relations to others, that are necessary to acting with integrity. On this side, we have capriciousness, wantonness, triviality, disintegration, weakness of will, self-deception, self-ignorance, mendacity, hypocrisy, and indifference. Although this category of excess dominates contemporary reflection on the nature of integrity, the first category also represents an enduring threat to integrity. Some people will be more prone to a certain set of practices or character traits that undermine integrity than others. The defeaters of integrity are therefore relative to individuals, and may even be relative to specific situations.

On this account of integrity as a virtue, integrity is much more difficult to achieve than is commonly thought. Virtue is a quality of character that one may have to a greater or lesser extent, in certain ways but not others, and in certain areas of one's life but not others. Integrity is not an all-or-nothing state. Rather, it is to make an "all things considered" judgment. Integrity as a virtue allows characterizing someone as having integrity even when we know, and even when *they* know, that in certain ways or about certain things, they lack integrity. Since a lack of integrity in one aspect of life does not necessarily mean there will be a lack in other aspects of life, a person could lack, for example, personal integrity while still having integrity in a number of other (restricted) areas of life, such as in intellectual or artistic pursuits, or, appropriate to our analysis, in areas of professional or career decisions.

In this line of reasoning, there can also certainly be conflict between types of integrity, particularly where the demands of a profession, say engineering, interfere with moral or personal integrity, say around war. Pursuit of one particular project can prevent one from balancing one's commitments. However, while different types of integrity can be sequestered from one another, integrity of one type is more likely to flourish in a context of greater integrity in various spheres of existence. The kind of virtues and skills that are developed in maintaining, say, intellectual integrity, are likely to be available for using to deal with the conflicts and temptations that threaten personal and moral integrity and conversely.

A conception of integrity as a virtue is also compatible with the existence of constraints on the content of the norms the person of integrity is committed to. Profound moral failure may be an independent defeater of integrity, just as hypocrisy, fanaticism, and the like are defeaters of integrity. One might judge as internal to our conception of virtue the idea that integrity is incompatible with major failures of moral imagination or moral courage, or with the maintenance of wholly unreasonable moral principles or opinions.[15]

Before returning to the question of what Georgina should do, it is important and relevant to discuss the ways in which the social structures shaping engineering, and contemporary liberal democratic societies at large, also impact our understanding of what integrity entails and how it can best be arrived at.

3.2   INTEGRITY AND SOCIAL STRUCTURE

As described above, integrity is often approached largely as a private or personal affair—albeit one with important implications in the public sphere. Yet, if integrity is as central and important a virtue as recent work on the topic suggests, then ideally the institutions, including forms of government

---

[15] On such a view, the Nazi could not, all things considered, be a regarded as person of integrity. The Nazi may be self-deceiver and a liar (which is likely), but even if he is not, his principles and his actions are not rationally defensible under any coherent moral view. And this latter fact may by itself justify the judgment that the Nazi, or the chemical engineer who works behind him, lacks the virtue of integrity.

and economic arrangements, that help shape our lives should also be structured in ways that promote integrity.

Unfortunately, less attention has been given to ways in which social, economic, and political structures and processes might affect personal integrity. Families, businesses, religions, and forms of government all affect personal integrity by promoting or undermining features essential to having or practicing integrity, or by aiding, abetting, or being inimical to the defeaters of integrity (such as self-deception).

As Susan Babbitt argues, an adequate account of personal integrity must:

> recognize that some social structures are of the wrong sort altogether for some individuals to be able to pursue personal integrity, and that questions about the moral nature of society often need to be asked first before questions about personal integrity can properly be raised…. The pursuit of adequate personal integrity often depends, not so much on understanding who one is and what one believes and is committed to, but rather understanding what one's society is and imagining what it could be. (1997, 118)

This approach leads us to question what kinds of society and what kinds of practice within a society are most conducive to personal integrity.

If society is structured in such a way that it undermines people's attempts at either knowing or acting upon their commitments, values, and desires, then such a structure is detrimental to the pursuit of integrity. And if integrity is connected to well being, then adverse social and political conditions are a threat—not merely an ultimate threat, but also an on-going, daily threat—to well being.

3.2.1 STRUCTURES OF ALIENATION

One 20th-century technical term for this mismatch is alienation. *Alienation* results when people are so confused or conflicted—are relentlessly exposed, for example, to the social manufacture of incompatible desires—that they take on roles they mistakenly believe they want or deceive themselves about wanting. Are political and social conditions in contemporary liberal democracies conducive both to acquiring the self-understanding necessary for integrity and, more generally, to the business of acting with integrity in day-to-day decision making?

Educational Structures

Historically, one of the governing ideals of liberal democratic societies is to provide its citizens not with the material objects they desire (e.g., washing machines, automobiles, and so on), but with what we might call "primary goods," such as freedom, and with the political, social, and cultural structures (e.g., laws, codes, institutions, practices, and so on) that facilitate citizens' capacity to obtain the material objects they desire. This is one reason education has always played a prominent role in discussions of liberal-democratic forms of life. Education is seen as a crucial structure in facilitating individuals' pursuit of both shared primary goods and their individually chosen material goods—perhaps, in the American parlance, these together could be called "the pursuit of happiness."

Such an instrumental view of education, however, is rather narrow and omits any role for inculcation of the means to choose among desired goods wisely. Integrity requires more than facilitation of an instrumental capacity to acquired desired objects. It requires the wisdom and self-knowledge to choose appropriate goods, worthwhile goals, and so on. It is, perhaps, hard to see existing mainstream educational structures playing a very significant role in this process, and harder still to imagine real institutions—institutions compatible with the demands and limitations of contemporary social and political life—that would.

Rather than facilitating the capacity of individuals to identify shared "primary goods" and act with discrimination around their individual desires, a host of theorists have argued that educational systems actually reinforce existing forms of social inequality and prop up the governments that sanction them. Italian social critic Antonio Gramsci made the links explicit when he wrote that "the State must be conceived as an 'educator.'" Through various apparatuses, including the university, "the State urges, incites, solicits, and 'punishes'...with moral implications" (1971, 247), and these implications can affect us in powerful ways.

In a similar vein, Louis Althusser (1971) argued that unequal societies, and particularly those structured predominantly by economic inequalities, are controlled by a combination of what he called repressive and ideological state apparatuses. *Repressive state apparatuses* include the military, the police, and the prison systems, which have the capacity to physically, violently repress dissent. *Ideological state apparatuses*, according to Althusser, are softer but arguably more effective, because they operate at the level of ideas and beliefs.

For Althusser, schools, churches, and related institutions help produce our very senses of self and, therefore, radically confine our opportunities for social action and substantive change, as well as the ways through which we can even imagine acting with integrity. We do not go as far as Althusser in imagining that opportunities for acting with integrity beyond these state apparatuses are so radically foreclosed, but we raise the issue to highlight how education itself influences our thinking in ways we may not fully recognize. We address this further in coming chapters.

Nevertheless, under the current structure of payment for much higher education—certainly in the U.S. and to a lesser degree elsewhere—students accrue substantial debt. Indebtedness forces a great many students—and more often those from working class or middle class backgrounds rather than the wealthy—into an immediate search for well-paying employment. Student debt and the threat of bad credit surely pressures many graduates into whatever sorts of employment they can find, discouraging their taking the time to find what could be a more fulfilling position that better accords with their beliefs.

Employment Structures

While existing mainstream educational structures may fail to facilitate the life of integrity, or else impart a specific shape or limit to what it can be, other structures are positively hostile to it. Arguably, and despite what might seem like overwhelming choice, job markets are structured by financial and other incentives, restricted opportunities, and economic rents. The result is that many people choose careers they do not really want, for which they are not really suited, and in order to make money as its own end, with little thought for the broader social good.

There are other, more straightforward ways in which social and cultural structures may be contrary to the pursuit of integrity as well. In professional life, people are frequently called upon to lie, bluff, or manipulate the truth in ways that directly or indirectly affect their individual integrity. The expectation that one "sells oneself" or "sells the company," for example, directly rewards both hypocrites and sycophants. And there are many kinds of employment assessments, reports, and application processes that foster both outward deception and self-deception.

Social Structures of Gender, Race, Class, ...

Social power structures—particularly those including by not limited to gender, race, and class—impact the experiences of all people in society, not merely those targeted by any particular "ism." Discrimination fueled by social power asymmetries has many faces. Only the most obvious such face is explicit and direct prejudicial treatment of one person by another—say, not hiring a person because of that person's social position and despite their qualifications. Explicit but indirect prejudice is another face of discrimination, but its damage is sometimes less apparent—here, think of sexist jokes told among male colleagues. An even less obvious, but in some respects the most insidious, face of discrimination is that which is weaved into the very fabric of our social structures. For example, structural racism is integral to many societies and manifests as cultural assumptions and stereotypes, extending from long histories of racial discrimination, and extending over a range of institutional arrangements. Importantly, such racism transcends any individual's or group's intentions—racist, anti-racist, or otherwise.

Race and class dimensions of discrimination are both highly relevant to our consideration of alienation and integrity, but we focus our attention here on gender as one example. Gender—the social meanings and power asymmetries associated with sexual and gender-based identity differences—is one of the most important social and cultural determinants in modern life. Common assumptions about the "nature" of men's and women's bodies, interests, and capacities are taken to justify and explain persistent social-role differentiation that results in inequalities in employment, familial obligation, child-care responsibilities, and other social practices. Despite these assumptions, these differences and the meanings associated with them can be shown to have important historical and cultural—rather than intrinsically biological—dimensions (Ortner, 1996; Butler, 1999; Stearns, 1979; Connell, 1995).

To this end, men who might otherwise prefer to dedicate themselves to childcare may feel themselves compelled to earn a high salary in order to be seen as a "real man," while women who seek professional careers can be attacked as selfish, as "masculine," as not being a "real woman," or as being a "bad mother" when balancing parenthood with a career. These issues can be particularly acute in engineering because of its historical masculinist culture—itself linked intimately with military institutions (Hacker, 1989)—but they clearly exist across society. Women engineers have sought to address these issues through professional societies, such as the Society for Women Engineers, founded in 1950, and a wide range of initiatives aimed at increasing women's representation among engineers. At the same time, they have clearly felt the pull of militarism as strongly as have many men (Guy, 2013a, 2013b).

Political-Economic Structures

Broad political-economic structures may also have a deleterious effect on our capacity to live with integrity, if in many small-scale but important ways. Contemporary society, driven as it often is by political-economic structures oriented insistently toward short-term gain, may itself be inimical to a life of integrity. While the effects of totalitarian regimes are certainly more extreme than those of liberal democracies, there is ample reason to be concerned about the latter as well. This is particularly so given the rise of neoliberal and neo-conservative ideology at the expense of a liberal democratic citizenry (Brown, 2006a, 2006b, 2003).

Those who are oppressed seem to be in a paradoxical relation to integrity. On the one hand, members of oppressed groups would seem to be deprived of the conditions for developing integrity: the freedom to make choices about how to act and think. As Babbitt (1997, 118) notes, one needs to be able to make choices in the first place order to develop the kinds of interests and concerns that are central to leading a life of integrity. On the other hand, oppressed people are often able to reflect with greater insight on political, economic, and social structures, because they do not benefit from them. They have no incentive to adopt self-deceptive or self-protective attitudes about circumstances of oppression or to see past them with convenient blindness. Oppressed groups, therefore, often have greater scope to confront social structures with integrity, and therefore to act upon this understanding with greater integrity. A capacity for reflection and understanding enables one to work toward integrity, even if it does not ensure that one achieves an ideal of integrity.

The kind of society that is likely to be more conducive to integrity is one that enables its members to develop and make use of their capacities for critical reflection, one that does not force them to take up particular roles because of their sex or race or for any other reason, and one that does not encourage them to betray each other, either in escaping prison or in advancing their careers. Social and political structures can be both hostile and favorable to the development of integrity, sometimes both at once. Such structures that pervasively impact engineers' ability to act with professional integrity are the subject of the following chapters.

We began this chapter with a discussion of a hypothetical Georgina to illustrate the complexity of the challenges and constraints that engineers (and others) are likely to face as they try to make career and life decisions with integrity. Like Georgina, each of us confronts a host of very immediate and personal constraints and enduring contradictions to leading the life we may aspire to. Integrity demands responding to such constraints thoughtfully, directly, neither pretending conflicts do not actually exist nor wishing away the trade-offs we are forced to make. In the case of engineers, integrity demands acknowledging the extent to which warfare and militarism pervade engineering as a field and as a way of understanding and engaging in the world. Individual engineers cannot singlehandedly change that reality, but they can, and should, we argue, make informed decisions that are responsive to it.

We hope that this treatment of integrity can help make sense of how to live in a world of complex social structures and constrained individual options, not by specifying the correct path for any individual engineering career but instead by mapping the terrain in a way that is honest to its complexity and contradictions. We turn now to the history and politics that have influenced the shape of this terrain and the very real problems, options, and opportunities on which engineers must creatively draw to navigate their careers with integrity.

CHAPTER 4

Historical Entwinements: From Colonial Conflicts to Cold War

President Dwight D. Eisenhower approached the microphone on January 17, 1961, as an outgoing President and an old soldier. His speech that day began with the mixed humility and pride of a soon-to-be elder statesman who understood not just the magnitude of his accomplishments, but also the scale of his failures. His words that day described an advancing militarism—an institutional formation he witnessed arising, and even actively nurtured throughout his career, but whose coming into being disquieted him.

Eisenhower had become president due to his life as a warrior. Educated at the U.S. Military Academy at West Point, Eisenhower led soldiers and established training regimens in the First World War, rising to the rank of a five-star general, and the Supreme Commander of Allied Forced during the Second World War. More than fifteen years after that war's conclusion, eight years of the presidency, and more than a decade into the Cold War, Eisenhower lamented that he had not done more to establish peace.

4.1 BIRTH OF THE MILITARY-INDUSTRIAL COMPLEX

The lasting note he struck in his final presidential speech was both prophetic and cautionary. Waging the Cold War had been the focus of his presidency, but rather than an invective against the dangers of Soviet totalitarianism, he warned of dangers posed by America's protectors: weapons industrialists and military planners, who came together in what he called a "military-industrial complex." These were men he knew well; he had been one of them for most of his life.

> This conjunction of an immense military establishment and a large arms industry is new in the American experience. The total influence—economic, political, even spiritual—is felt in every city, every State house, every office of the Federal government. We recognize the imperative need for this development. Yet we must not fail to comprehend its grave implications. Our toil, resources, and livelihoods are all involved; so is the very structure of our society.

The arms industries, he warned, threatened to reorient the entirety of the United States' political and economic—and even spiritual—foundations. Coming from a career soldier-politician, this was no small matter to address.

In the councils of government, we must guard against the acquisition of unwarranted influence, whether sought or unsought, by the military-industrial complex. The potential for the disastrous rise of misplaced power exists and will persist.

Eisenhower was concerned about undue influence in government, and about the potential for those in the armaments and related industries to set priorities for governmental policy, at home and abroad. Indeed, he warned how the military-industrial complex threatened the world beyond government, impacting the structures of civil society, of citizenship, and of education—the foundations of a free citizenry.

He was especially concerned about how new modes of military-based research could skew intellectual inquiry, particularly university-based research. Academic inquiry in a liberal democratic society—but one in which armed forces played a key role—demanded a delicate balance. "Only an alert and knowledgeable citizenry," Eisenhower warned, "can compel the proper meshing of the huge industrial and military machinery of defense with our peaceful methods and goals, so that security and liberty may prosper together." One should not outweigh the other.

Today, the solitary inventor, tinkering in his shop, has been overshadowed by task forces of scientists in laboratories and testing fields and engineers in large, bureaucratic organizations. In the same fashion, the free university, historically the fountainhead of free ideas and scientific discovery, has experienced a revolution in the conduct of research. Partly because of the huge costs involved, a government contract becomes virtually a substitute for intellectual curiosity.

The prospect of domination of the nation's scholars by Federal employment, project allocations, and the power of money is ever present—and is gravely to be regarded. Yet, in holding scientific research and discovery in respect, as we should, we must also be alert to the equal and opposite danger that public policy could itself become the captive of a scientific-technological elite. It is the task of statesmanship to mold, to balance, and to integrate these and other forces, new and old, within the principles of our democratic system—ever aiming toward the supreme goals of our free society.

It should be said that Eisenhower was no dove or peacenik; he oversaw numerous wars and supported a powerful military establishment. Indeed, for much of his career he urged closer relationships between the military and profit-driven corporations. In 1928 and 1932 papers before the Army War College and the Army Industrial College, Eisenhower explicitly called for closer military and industrial peacetime collaboration (Roland, 2001, 3). As president, he championed the expansion of capitalist markets across the planet, and suppression of governments that would limit or regulate capitalism in their own nations.

Yet the priorities and dangers Eisenhower emphasized in his farewell address were in many respects prophetic. The impact of militarized research eclipsing other forms of knowledge production has been profound. The critique was taken a step further by another American statesman. Senator William Fulbright took Eisenhower's challenge and extended it, warning of a "military-industrial-academic" complex, and its potential of limiting the university itself. "In lending itself too much to the purposes of government," Fulbright warned, "a university fails in its higher purpose" (Fulbright, 1970, as quoted in Giroux, 2007, 15).

While Eisenhower had been working his way through the U.S. military establishment and training soldiers to fight in the First World War, a social critic named Randolph Bourne warned of the dangers war posed to democracies. The horrors of the First World War, in which the full weight of modern-engineering technologies and the productive might of industrial capitalism were put to the ends of human destruction, were fresh at hand. Moreover, Bourne had seen the war prompt a massive wave of political repression in the U.S., which vilified ethnic minorities as "un-American." Even modest questioning of the war effort was attacked as unpatriotic, and systematically crushed in a widespread anti-Communist Red Scare.

In an essay entitled "The State" (1918), Bourne investigated these and other facts of the wartime world. In that essay, he concluded, provocatively, that war is "the health of the state." Bourne was not a happy advocate of such a state, and, unlike the young Eisenhower, he did not champion its expansion. Rather, as an internationalist and political dissident, Bourne feared its wrath and its anti-democratic tendencies. Bourne argued that war forced citizens to disregard domestic politics in favor of a transcendent nationalism. The states that worried Bourne were invested in territorial acquisition, the control of new resources and markets—this was the stuff of his day's geopolitical struggle.

But politicians were also interested in controlling their domestic populations, and securing political power against those who might challenge it—even by democratic means. They relied upon not just on military technologies to win battles abroad, but also policies to secure consent among their own citizens. The danger Bourne identified was not that democracies would be overrun by militarized adversaries, but rather that militarization had the corrosive power to undermine democracy from within. He anticipated both Eisenhower's and Fulbright's warnings by some 40 years.

This chapter takes up the challenge set by Bourne, Eisenhower, and Fulbright to investigate the connected histories of engineering, engineering education, and warfare. It returns to the periods well before the Cold War, and examines nationalism, warfare, and scientific inquiry in the production of society and the forms of knowledge—including engineering knowledge—deemed most socially important.

The relationship between warfare and technological development, and thus engineering, is a long one. The issues emerged first and are perhaps starkest in the hard sciences and engineering. But as concepts of militarization and of political power have expanded, the implications are increasingly important across the entirety of the University.

4.2   THE MILITARY-INDUSTRIAL-ACADEMIC COMPLEX

In the chapter that follows, we define a military-industrial complex as "an informal and changing coalition of groups with vested psychological, moral, and material interests in the continuous development and maintenance of high levels of weaponry, in preservation of colonial markets and in military-strategic conceptions of international affairs" (Pilisuk and Hayden, 1965, 103, as quoted in Pursell, 1972, ix). The roots of these connections emerged in preindustrial societies, but they have changed over time, and particularly in moments when global geopolitical forces have been in flux. Consequently, the chapter will describe some of the longer tendencies and connections between engineering, politics, academia, and warfare.

Technologies contribute to historical change, yet at the same time, technologies only become relevant and embraced when their "proper" historical moment comes. In other words, technologies require the historical confluence of multiple political, social, cultural, and environmental forces, which allow the technologies to "stick," or else be shed. New military technologies in the 1890s helped facilitate modern, industrially driven colonialism and the accrual of colonized lands and resources into European imperial treasuries. But when colonial powers fought against each other, those same forces sowed the seeds of their own destruction. They came to full and deadly bloom in the Second World War.

Just a few years later, the Cold War between the Soviet Union and the U.S. gave rise to new, neocolonial practices where military victory was tied less to controlling land than to exerting influence and profiting from resource extraction. As the economy was increasingly globalized, and people and goods traveled more and more freely across national borders, these tendencies intensified. Today, in the 21st century, military wares have come to embrace a far more thorough range of technologies, from biotechnology and pharmacology to informational controls with the older means of applying force. The case studies we examine are intended to illustrate some key features of the connections between military research and the broader social order in the modern world. It is far from an exhaustive discussion.

4.2.1   LONG HISTORIES, GLOBAL HISTORIES

As described in Chapter 1, engineering and militarism have deeply intertwined histories that date to the medieval period and before. Fortresses were built not just to repel invaders, but to guarantee and protect the social inequality of feudalism, securing feudal lords from the danger that the serfs beneath them might raise arms. The military engineer, according to theorist Paul Virilio, became "a war entrepreneur...at the origin of the State armies and later of the military industrial complex" (1977, 37). The connections between technical design and sovereign powers—be they feudal estates, kingdoms, empires, or later, modern nations—has a deep historical past.

If the historical foundation for military and political might is deep, its geographic reach is also wide. Nevertheless, in the section that follows we concentrate on European trajectories, and the histories they spawned in the wake of 19th-century industrialization and the modern world that followed. Though many observers associate a military-industrial-academic complex with the post-WWII U.S.—and this makes sense, given Eisenhower's centrality to framing the concept—the term has broader application across the modern world.

British Precedents

The British version of a military-industrial complex might be traced from the Anglo-Saxon period, when the King's private army was provisioned by weapons manufacturers and weapons were then stored and maintained in a royal arsenal. Scholars have found related military-political complexes since then, but a more salient example for our purposes was the naval arms race between Britain and Germany at the turn of the 20th century. It was heightened in the period surrounding the Berlin Agreement of 1890, in which the great European powers agreed on how to partition much of the Southern hemisphere—Africa, Asia, Latin America—into Northern territorial colonies. Suffice to say that few representatives from those continents were allowed to offer their opinions.

As the imperial powers vied for influence and access to global markets, Britain and Germany understood their navies as fundamental to their economic and military competition. On the British side, naval officers and industrialists cooperated to influence national policy. They saw German naval expansion as a threat to their imperial interests, and persuaded the people and Parliament that Britain needed a massive increase in naval funding. Shipbuilders, for their part, were happy to accept that money.

As it turned out, British shipbuilders and other armaments manufacturers had deep and abiding personal relationships with their counterpart naval officers. They all tended to be upper class, had been educated in the same elite universities, and shared a common belief that the maintenance and expansion of British Empire—controlling the resources and subordinating the peoples of much of Africa and Asia, while excluding other Europeans from those markets—was a fundamental good (Roland, 2001, 6–7; Higham, 1981).

The military and political vision is nicely captured in the term "gunboat diplomacy." Colonized governments would agree to terms that favored the colonizers when faced with overwhelming military—and naval—force. Literally, when colonizers sent big guns on ships, local governments would capitulate to their economic and political demands.

German Precedents

The Germans, for their part, had a somewhat different experience in the late 19th and early 20th centuries, though they, too, wanted their own ships and guns. The disparate German states were united under Otto von Bismarck in the late-19th century. Once drawn together, Germans had grand aspirations of territorial expansion across central Europe and beyond. Its political leadership and driving vision came from the Prussian tradition of extensive militarism under a centrally organized and directed state, and in which the army came to represent all that was considered "best" about Germany.

In the imperial age, previous desires for continental expansion seemed inadequate, and Germans soon believed that competition with other major imperial powers—especially the British—required the access to global markets that overseas colonies provided. This, in turn, required a strong navy. Indeed, shipbuilding led to an alliance between Germany's industrialists and expansionist nationalists. Moreover, and not least, expanding industrial labor in the shipyards promised to reduce the perennial problem of domestic class conflict by promising good jobs and high wages to German workers. This sort of accord, linking high wage jobs at home to militarism overseas, would have echoes in the postwar United States.

Despite these tendencies, the First World War caught the German military establishment somewhat unawares, because its planners had anticipated a short war. State planners quickly determined that they needed more direct oversight over munitions and military preparedness, and that they could not leave such preparations to the working of the market. Though the basic capitalist foundation of the economy remained intact, there was a pronounced "melting" of the lines of authority between government and business, more than had been the case in Britain (Homze 1981, 61). German weapons and other manufacturers typically did quite well for themselves from the bargain, but throughout the First World War—and even through the 1930s and the Nazi period—weapons producers generally found themselves politically subordinated to the dictates of state planners. Only rarely did this dampen profits.

By the 1930s, firms that politically and materially supported the Nazis—and were willing to purge their Jewish employees, such as chemical giant I.G. Farben—could do quite well for themselves. I.G. Farben and its subsidiaries would innovate industrial techniques vital to the German war effort, including producing synthetic rubber, refining fuel, and developing poison gasses, including Zyklon B. I.G. Farben's materials went into virtually every facet of the Nazi war machine, and it drew profits from them too—not least from the slave labor and death camps at Auschwitz (Jeffreys, 2008; Homze, 1981).

The Nazis are of course an extreme case of engineering and industrial development wedded to militarized geopolitical expansion, but tendencies existed among more moderate nations, too, and in periods of relative calm, rather than just in wartime. Indeed, evidence from Sweden and Japan in the late-20th century similarly underscores the ways that nationalist regimes and military production can collaborate (Ikegami-Andersson, 1992). Neither arms races nor superpower status are necessary for producing relationships between firms and states that impact political configurations and shape the possibilities of political action and academic knowledge across a range of fields. There is not one kind of military-industrial complex; there are many.

4.3   SOCIAL HISTORY OF THE MILITARY-INDUSTRIAL-ACADEMIC COMPLEX

Dwight Eisenhower spent most of his adult life trying to make the U.S. military into an effective fighting force, coordinating the massive logistical and infrastructural needs of industrial war-fighting. The relationship between military development and industrial growth, with all of the attendant engineering and design problems and solutions, has implication far beyond the drafting table, the boardroom, or the military planners' desks. Indeed, if we take the case of the U.S. from the Second World War to the present, we can see how thoroughly the tenets of military production penetrated the social order, from popular culture to architecture to people's understanding of their racial, gender, and class identities, among others.

Beginning with World War II and persisting through the Cold War, military spending was the literal foundation of the U.S. economy (Chafe, 2010). The policy was solidified in what we might call Military Keynesianism. This is the economic idea that governmental spending would bolster geopolitical security while stimulating domestic consumption, driving everything from road-building programs to factory openings. When workers in defense and allied industries were paid high wages, they might spend that income on new washing machines, automobiles, and homes, which would then create additional well-paying jobs in manufacturing and construction.

This policy was predicated on a political bargain among the federal government, large corporations, and the centrist trade unions. Corporations agreed to pay high wages with good benefits; trade unions would assure corporations that their members would not go on strike. Domestically, the U.S. government would manage those industrial relations, while fighting overseas to guarantee that U.S.-dominated capitalism prevailed over Soviet communism and other economies across the so-called third world. And those lands brought within the orbit of the U.S. economy (known as the "free world") would be more likely to buy American-made washing machines and automobiles, while sending their own raw materials to the United States.

All of this required massive building, massive spending, and massive engineering, which were directed by states and governments, but also by private corporations—and the best work was done when their interests aligned, as in World War II and the Cold War (Lipsitz, 1994). The genius and danger of a military-industrial complex, its critics worried, was that it tied national economic well-being to ever-expanding weapons programs.

In the U.S., a shift in weapons contracts meant the loss of jobs, and the massive dislocation that unemployment wrought. Perversely, then, the wellbeing of American families was founded on machines and tools that could—and in hot wars, did—destroy other peoples' families. Though the Cold War was understood as a battle between the Communist East and capitalist West, in point of fact the actual, hot wars of the era were fought in the global South. Indeed, the broader context of the Cold War was within a fundamental reshaping of the globe's resources and nations.

The people of Africa, Asia, and Latin America had long chafed against the European colonial impositions written into the 1890 Berlin Agreement and underwritten by gunboat diplomacy. The Second World War destroyed that global political and economic system, and anti-colonial movements took root around the southern hemisphere. The Cold War emerged as a new struggle between the Soviet Union and the U.S. to reorder the globe, while non-aligned and newly independent nations struggled to control their own resources.

4.3.1   FROM MILITARY TECHNOLOGIES TO SOCIO-ECONOMIC PRACTICES

The Second World War revolutionized not just military technologies but also social practices. The war was fought in Europe and the Pacific, and the battle ravaged economies across these regions. Few German, French, British, Japanese, or Russian factories escaped bombing. But the American industrial and agricultural economies remained unscathed, and the U.S. was poised to provide the aircraft, ships, weapons, and foodstuffs necessary for the global wartime economy.

Once the U.S. entered the war, demand for soldiers meant that new opportunities arose for black and Latino men and all the women workers who had been deemed marginal to the U.S. industrial economy, and this provided the chance for their democratic social advancement. When Asa Philip Randolph, African-American leader of the Brotherhood of Sleeping Car Porters, threatened a March on Washington in 1941, President Franklin Delano Roosevelt agreed to pass an Executive Order guaranteeing fair employment in defense industries. As a result, black men could be hired in jobs that had previously been reserved for whites-only.

Moreover, white women and women of color began to work in higher-paying jobs, rather than being relegated to unpaid housework or low-wage domestic service. Wartime demands for labor, coupled with the ideology of a war against racist fascism, allowed for new social possibilities. For many women and for African Americans, the Second World War's military-industrial complex provided new opportunities for social inclusion in the U.S. Yet the domestic cost for Japanese Americans was high—they were imprisoned as "enemy aliens" in Internment Camps without evidence and without trial.

Understandably, given the use of atomic weapons, U.S. soldiers and politicians believed that evermore sophisticated weapons and logistical systems won World War II for the Allies, and were committed to protecting U.S. military and technological dominance. (The American memory of atomic victory in the Pacific also helped to downplay the importance of Soviet responsibility for the Allied victory in Europe, an amnesia that served Cold War aims.) As a result, and because of the many kinds of specialized knowledge these technical systems demanded, universities were also unexpected beneficiaries of industrialized warfare. Funding was especially forthcoming in the physical sciences: physics, materials science, etc. By the end of the Second World War the Massachusetts Institute of Technology (MIT) topped the list of those receiving government largesse, with around $117 million in R&D funding. Caltech received $83 million, Harvard and Columbia received around $30 million each (Leslie, 1993, 14).

After the Second World War, links between the military and universities continued to gain strength. The Navy became intimately linked with John Hopkins University's Applied Physics Laboratories, the Army joined with Caltech's Jet Propulsion Laboratory, the Atomic Energy Commission joined hands with UC Berkeley's Los Alamos Weapons Laboratories. Despite considerable cuts in military spending with the end of the war, funding for research and development remained remarkably stable. It would quickly increase with each round of postwar international crisis: in response to the Berlin crisis, the Soviet Atomic bomb tests, and the success of the Chinese revolution (Leslie, 1993, 8).

The Korean War proved to be the moment when universities became full partners in the military industrial complex. "Virtually overnight, defense R&D appropriations doubled, to $1.3 billion" (Leslie, 1993, 8). Military money poured into academic and industrial research centers. Existing contracts for applied and classified research were expanded, and entirely new university laboratories took shape. MIT's Lincoln Laboratories would concentrate on air defense; Berkeley's Lawrence Livermore Laboratories focused on nuclear weapons; Stanford's Applied Electronics Laboratories would develop electronic communications and countermeasures, which led to the birth of California's Silicon Valley (Leslie, 1993, 8).

Grant and research money flowed to programs whose research directly led to weapons or military-useful technologies, and membership in those teams was considered an important feature of academic and professional success. As senior researchers won grants and accolades, younger scholars became acculturated to that research model. Graduate students were quick to pick up on who got the big grants, who was promoted, and why. More often than not, they realized that the most successful career paths went by way of major defense contracts, advisory positions for the military, or consulting jobs in the defense industry (Leslie, 1993, 148). As physicist S.S. Schweber reflected, "These outstanding scientists helped create a social reality in which working on military problems was an accepted norm for scientists at universities; more than that, their involvement suggested that one aspect of being the very best was participation in defense projects" (Schweber, 1988, quoted in Leslie, 1993, 8).

4.3.2   FROM SOCIO-ECONOMIC PRACTICES TO TECHNOSCIENTIFIC RESEARCH

U.S. politicians and military planners understood the Cold War to be a struggle between freedom and authoritarianism, which would require full social mobilization. Despite structural, political, and economic differences between the U.S. and the Soviet Union, their military research institutions revealed some surprising similarities. Direct patronage from the military powerfully redirected the nature and practice of scientific and engineering research, defining both the "problems" to be solved and priorities for ongoing academic work (Leslie, 1993, 9). In many respects, this military emphasis came to define what it meant to be an engineer or a scientist in the Cold War period, on either side of the Iron Curtain.

The production of nuclear weapons is perhaps emblematic of a new kind of "big science" and the growing power—military and bureaucratic—of the linkages between academic research and narrowly defined national security interests. The irony in nuclear weapons research was that it produced its own profound insecurities of national, and even global, annihilation. U.S. and Soviet emphases on nuclear security, via technological advancements in warhead delivery or explosive power (which increased by many thousandfold between 1945 and 1995) continuously revealed their own spiraling susceptibility to total destruction. Nevertheless, despite the omnipresent threat of destruction, the nuclear economy has been profoundly successful in hiding itself from public scrutiny (Masco, 2006, 3–4, 6).

The production of nuclear weapons required a profound reorientation of national priorities. Early in the Manhattan Project, physicist Niels Bohr quipped that the entire U.S. would have to turn into a factory in order to produce the uranium to make a nuclear bomb. A half-century later, he was largely proven right, and the nuclear infrastructure expanded to a national scale. Between 1940 and 1996, the U.S. spent more than $5.8 trillion to produce seventy thousand nuclear weapons (Masco, 2006, 18).

Developing and maintaining the U.S. nuclear arsenal was perhaps the largest industrial enterprise in world history, with attendant industries in building roads, developing emergency bureaucratic structures, and countless redundant emergency systems. The permanent and omnipresent threat of nuclear war gave rise to an infrastructure of early warning systems; constantly circling, nuclear-armed bombers; and "always-on" computers whose networks and surveillance systems enwrapped the world (Masco, 2006, 25; Edwards, 1996).

4.3.3 BIG SOCIAL SCIENCE

If "big science" was a crucial feature of Cold War physical sciences, it also had an analogue in the social sciences; we would be remiss to ignore how academic social science was brought into the Cold War as well. Planners understood with great clarity that the Cold War would be fought at the level of idea and ideologies, and not just with bombs, bullets, and other kinetic weapons. Both the Soviets and the U.S. tried to convince the formerly colonized peoples of Africa, Asia, and Latin America that their empire was their best option for the future. Each empire sought to use the social sciences to bring previously marginal economies and regions into their relative spheres of influence (Simpson, 1998, xvi).

The U.S. began a concerted program to research the best ways to communicate and "sell" its geopolitical and economic vision to formerly colonized peoples. On U.S. campuses, this included the institutionalization of such fields as area studies, development studies, operations research, and communications research, among others. These "big social science" programs paralleled their cousins in the physical sciences. At least six of the most important postwar communication studies programs earned more than seventy-five percent of their annual budgets from government funding. Columbia University's Bureau of Applied Social Research, Princeton University's Institute for International Social Research, and MIT's Center for International Studies were among the institutions engaged in de facto psychological warfare operations, up to and including studies in the psychology of torture (Simpson, 1998, xii–xiii).

Programs in development studies provided a very particular intellectual background for the ongoing debates around global economics in the early Cold War years, and this intellectual foundation was predicated on the necessity and fundamental good of opening the so-called Third World to the penetration of U.S.-based markets while denying access to the Soviets. Theorists did not shy away from planning repression for those regimes that had different ideas for their own economies.

In a 1954 report to CIA chief Allen Dulles, MIT-based Max Millikan and esteemed development economist Walt W. Rostow argued that the U.S. should even subordinate respect for democratic processes in Asia and Latin America if it meant that pro-U.S. regimes would take power. Indeed, believing that capitalist growth was a precondition for democracy, they argued that capitalism must logically come first, and that some political repression might be regrettable but necessary to aid "traditional" societies in what they saw as a linear transition to capitalist modernity. Economic incentives and internal security measures would be necessary, they wrote, to "create an environment in which societies which directly or indirectly menace ours will not evolve" (reproduced in Simpson, 1998, 39–55, esp. 41; see also Rostow, 1960). Anthropologists, economists, and communications studies scholars, no less than atomic scientists or electronic engineers, would work to make that happen.

The Cold War ended, thankfully, not with nuclear cataclysm, but with the effective surrender of the Soviet Union. Radically centralized state planning proved less dynamic and less economically efficient than the messy and diffuse nature of U.S. political systems and attendant military spending. Some intellectuals saw the conclusion of the Cold War as the "end of history" (Fukuyama, 1992) and as ushering in a utopian victory of Western-style liberal democracies and the permanent expansion of capitalist markets.

But history, clearly, has continued. Economic, racial, and gender inequities persist within and across nations and national borders, and in many cases have worsened. Geopolitical struggles among nation-states remain. The current period of advanced capitalism and late modernity has seen the undoing of Cold War geopolitical and economic certainties, and new kinds of political, economic, and religious allegiance (Harvey, 1990, 2007). But these, too, had precedents during the Cold War, and governments experimented with new strategies to control domestic populations. It is to the new and attendant forms of state-making, engineering, and war that we now turn.

CHAPTER 5

Historical Entwinements, Post-Cold War

Nineteenth-century global colonialism had been underwritten by the practice of gunboat diplomacy, but the second half of the 20th century saw an expanded range of scholarly fields in the application of national geopolitical goals. This reflected an understanding that the application of kinetic force and physical violence had serious strategic limitations in war-fighting, and reflected a relatively narrow vision of how struggles would be waged into the future.

This chapter addresses the historical contexts of this most recent turn in connections between engineering and warfare, and the move beyond traditional conceptions of the application of physical sciences and kinetic weapons. It concentrates on the application of "non-lethal" force as a strategic (rather than ethical) concern, and the ways in which the Defense Advanced Research Projects Agency (DARPA), perhaps the most sophisticated and certainly the best-funded defense research institute in the world, reflects these new understandings of what warfare might mean.

5.1 "SOFT KILL" WEAPONS RESEARCH

As the world entered the last quarter of the 20th century, lines between enemy and friendly territories grew hazier. The period coincided with the rapid expansion of global markets, which has come to be called the era of globalization, advanced capitalism, or late modernity. The use of military technologies evolved from their initial impetus as weapons pointed "outward" toward external enemies, to—perhaps reprising the feudal fortress's insurance against serfs—now also point "inward," with many domestic applications.

During the United States' war in Vietnam (having assumed the mantle of colonial control from the French in 1954), and in its aftermath, many military personnel and technologies were brought to bear on non-combatant populations. To some extent this reflected the fears of a "postcolonial" world and the demise of clearly defined boundaries between European or U.S.' colonies in Africa, Asia, and Latin America. It also reflected a shift in military tactics toward counterinsurgency doctrine rather than reliance on overwhelming force. At the same time, migrants from former colonies made their way to European or North American metropoles. By the late 1960s, political protesters in the U.S. and Europe came to embrace the symbols and anticolonial ideologies espoused in the Third World.

In 1965, the National Guard was called out in Watts, Los Angeles, when thousands took to the street to protest police violence against African Americans. The Rand Corporation, a think-tank that dedicated much time and money to counterinsurgency in Vietnam, brought its expertise to suppressing rebellion in Los Angeles. Similarly, the Federal Bureau of Investigation's Counter-intelligence Program, known as COINTELPRO, made an explicit goal of crushing the radical civil and economic rights activists in the Black Panther Party.

Soon thereafter, new police units, modeled explicitly on elite military forces, took shape. Employing cutting edge weaponry, these Special Weapons and Tactics (SWAT) teams proliferated across the country. Such programs often targeted urban areas, but by the 1980s, similar processes took shape to militarize the U.S.-Mexico border against the arrival of undocumented Latin American migrants. Here, too, new military technologies grew and took root within the U.S., with analogues arising on Europe's borders (Parenti, 1999; Dunn, 1995).

5.1.1   THE RISE OF NON-LETHAL WEAPONRY

In the era of the global economy, the lines between supposed external and internal enemies blurred. Consequently, military and policing imperatives shifted. Despite numerous and ongoing advances in lethal weaponry, domestic police forces sought additional means of control. In the wake of the riots and protest movements of the 1960s and 1970s, from Civil Rights and feminism to student movements and queer rights, the police looked to new, non-lethal weapons programs for crowd control and to tamp down protest. According to their advocates, non-lethal weapons are defined as "[w]eapons that disrupt, destroy or otherwise degrade functioning of threat materiel or personnel without crossing the 'death barrier.'"[16] These are among the rapid growth industries today.

The desire to make weapons that inflict pain to the point of incapacitation—but without the risk of death—also has roots in the Cold War. In 1949, the American military researcher L. Wilson Greene wrote a classified report entitled "Psychochemical Warfare: A New Concept of War." Soviet researchers engaged in similar projects, though we know less of their findings. Nevertheless, Greene and his colleagues—researchers in pharmaceutical industries, in psychiatry and medicine, and in the military—sought ways to win wars without death. "Throughout recorded history," Greene lamented, "wars have been characterized by death, human misery, and the destruction of property; each major conflict more catastrophic than the one preceding it." Chemistry and pharmacology promised a new dimension, and a new military frontier.

Earlier chemists had of course developed poison gasses, but according to Greene, a less cruel day beckoned. "I am convinced that it is possible, by means of the techniques of psychochemical warfare, to conquer an enemy without the wholesale killing of his people or the mass destruction of

---

[16] John Alexander, "Non-Lethal Defense" (briefing), Los Alamos National Laboratory, 1993, quoted in Charles Swett, "Strategic Assessment: Non-Lethal Weapons," (Office of the Assistant Secretary of Defense for Special Operations and Low-Intensity Conflict, November 9 1993), p. 4. We note that the trajectory described by these military theorists—and that we discuss throughout the chapter—aligns with what Michel Foucault (1979, 1990) saw as a historical shift from older forms of violent rule to more modern and subtler means, which he called *disciplinary* and later *biopolitical* forms of power.

his property." While Greene and his colleagues aimed to debilitate enemies by chemically inducing seizures, fear, panic, hysteria, and even suicidal mania, a utopian ethic of non-harm remained one of their goals. James Ketchum was deeply involved in the chemical testing, and would later recall "I was working on a noble cause…. The purpose of this research was to find something that would be an alternative to bombs and bullets" (Khatchadourian. 2012, 50-51, 57).

By the late 1960s, much of the research that Ketchum conducted fell apart. The military application of the projects was questioned and participating physicians could no longer ignore their ethical concerns over the injuries done to the people on whom they experimented. Yet in recent years, there seems to have been a return to such "non-lethal" weaponry. In some part the resurgence came in the wake of the social protest movements of the middle and late-20th century, but it has come to fuller realization as U.S. geopolitics in the 21st century have tried to emphasize "soft power." For example, in the midst of the U.S. and allied forces' effort to occupy Iraq and maintain control over a fractious population in Afghanistan in the name of the so-called Global War on Terror, their political ambitions have been set back time and again by unintentional civilian deaths.

U.S. military strategy has sought to minimize the number of non-combatant deaths, but this is a perennial problem of war—and one quite obviously worse for non-combatants than for military planners—without a technical or tactical solution. With a media apparatus that can make a single innocent death into a powerful symbolic event, the U.S. has tried to improve upon non-lethal technologies, especially for crowd control (Arike, 2010).[17] Controlling living people through the calibrated, technologically mediated infliction of pain-without-death, then, might be understood as a new way of waging war.

Older military technologies concentrated on exercising force through traumatic kinetic force: Bodies would be torn apart, buildings would be crushed. This has not disappeared. But the newer means of waging war understand and apply force in different ways. An examination of DARPA's newer programs, as well as systems with increasing application by police forces against their own citizens rather than external populations, suggests that new technologies and weapons systems may have a different relationship to bodies. Weapons can focus on the molecular structure of bodies, the function of electricity in musculature, the senses of smell and of hearing and of taste. Drawing on a number of desired medical or physiological effects—such as pain, lethargy, or disorientation—contemporary "non-lethal" weapons are essentially "reverse engineered" to produce those disabling effects on individuals or groups (Arike, 2010, 45 note 14).

5.1.2   CIVILIAN CROWD CONTROL

There are of course a dizzying range of such new weapons available, and many of these beggar the imagination, save for those steeped in what until recently was considered science fiction. The U.S.

---

[17] Unless otherwise cited, the discussion of "soft-kill" weapons that follows draws from Arike's (2010) incisive article.

Pentagon prefers the term "non-lethal," but manufactures refer to them by terms like "soft kill," "less-lethal," "compliance," and "low collateral damage." They are not intended for use against combatants—soldiers or militants—but rather against unarmed or primitively armed civilians. In other words, they are intended for crowd control.

A joint Pentagon and U.S. Justice Department report warned in 1997 that "Even the lawful application of force can be misrepresented to or misunderstood by the public," and therefore, "[m]ore than ever, the police and the military must be highly discreet when applying force" (Arike, 2010, 39). The report's authors felt that their task—of attempting to control civilians occupying public space, often as a means of expressing otherwise denied political opinions—was hampered by outdated tools. As a U.S. National Advisory Commission on Civil Disorders noted in its 1968 report: "The police who faced the New York riot of 1863 were equipped with two weapons: a wooden stick and a gun. For the most part, the police faced with urban disorders last summer had to rely on two weapons: a wooden stick and a gun" (Arike, 2010, 41).

These options seemed inadequate to controlling urban protestors. Batons would do too little to stop them, but the use of firearms was not only literal overkill, but would be counterproductive, making martyrs of the protestors and inciting further unrest. Eventually, Congress funded an Omnibus Crime Control and Safe Streets Act, which funded a "veritable Manhattan Project" for researching new policing technologies and tactics. Scientists, engineers, and military planners came together yet again. In 1971, the National Science Foundation researched the problem, and its publication, *Nonlethal Weapons for Law Enforcement: Research Needs and Priorities*, would "identify areas in which scientific research can help solve social problems" (Arike, 2010, 42). Not the social problems of racism or income inequality (which were the underlying causes of urban protest in the 1960s), but rather of the disorder that these problems provoked.

5.2   NON-LETHAL WEAPONS RESEARCH COMES OF AGE

In the early 1990s, then-Secretary of Defense Dick Cheney commissioned a "Nonlethal Strategy Group," whose purpose was to "integrate available capabilities into military doctrine and inventories, and guide future investment into research and development of promising capabilities" (Swett, 1993, 1). Researchers at DARPA were undertaking investigations into Nonlethal Weapons (NLW) possibilities. National research labs and various other private contractors had been "pursuing the technology" in hopes of the sorts of contracts that might eventuate. The Research and Development arm of the Department of Justice also "expressed strong interest in taking advantage of the [Department of Defense's] efforts" (Swett, 1993, 3).

By 1992, the U.S. Army had drafted desired operational components of these weapons and their "disabling measures capabilities" (Swett, 1993, 2). They would:

- Impair human capabilities
    - Temporarily dazzle or overcome human operators with intense light
    - Disperse crowds using transient-effect generators which produce a frequency or sound to temporarily immobilize of disorient humans
    - Calm people or put them to sleep
- Defeat materiel systems
    - Blind optical sensors and targeting devices
    - Destroy or inactivate electronics, including electronic ignitions, detonators, communications, and radars
    - Cause vehicles to stop or keep aircraft from flying
    - Ignite/destroy reactive armor
    - Cause computer driven systems to malfunction or induce operating errors
- Attack strategic and tactical materiel support systems
    - Weaken or change fuels and metals
    - Contaminate or plug water pipeline
    - Defeat modern materials (i.e., composites, polymers, alloys) (Swett, 1993, 2).

Charles Swett, Assistant for Strategic Assessment and Defense-Department analyst, expressed great excitement at the military potential of these weapons. Moreover, in his 1993 report, he beamed that "a large conference on NLWs will be held at the Applied Physics Laboratory of Johns Hopkins University." No less a personage than the Attorney General would address the civilian and military luminaries in attendance, who would, no doubt, share concerns that increasingly overlapped. Conference participants would include a former Army Chief of Staff, "several active duty three-star flag and general officers," and, perhaps most illustrious of all, the theoretical physicist and hard-liner nuclear Cold Warrior, Edward Teller (Swett, 1993, 3).[18]

The prospects of urban warfare against civilians or poorly armed insurgents seemed highly likely to military planners around the world, and research on non-lethal compliance weapons was ongoing. By the late 1990s, NATO, Israel, the United Kingdom, and Canada all undertook similar research programs, liaising with the field's leaders in the U.S. In the early-21st century, China and Russia began their own programs on non-lethal weaponry (Arike, 2010, 44).

[18] For biographical information on Teller, see: http://www.atomicarchive.com/Bios/Teller.shtml, accessed April 13, 2012.

5.2.1 SECOND-GENERATION SOFT-KILL WEAPONRY

Recently, researchers have shifted focus to "second generation" soft-kill weapons, pressing beyond the rubber bullets or pepper spray of previous years. According to one investigator, "the trend is now away from chemical and 'kinetic' weapons that rely on physical trauma and toward post-kinetic weapons that, as researchers put it, 'induce behavioral modification' more discreetly" (Arike 2010, 45). Despite some researchers' concerns over potential neural damage, microwave-based weapons—"in which short microwave pulses rapidly heat tissue, causing a shockwave inside the skull that can be detected by the ears"—are gaining traction (Hambling, 2008).

These sorts of disablements—which attack the body at the level of the inner ear or its electrochemical systems—are intended to be non-lethal, though as many cases of TASER-induced deaths can attest, this is not always the case.[19] Nevertheless, TASERs have grown tremendously in popularity with police forces around the world, especially because they are understood as a safe alternative to firearms. They have become something of an "all-purpose tool for what police call 'pain compliance'" (Arike, 2010, 46).

Medical researchers linked to Pennsylvania State University's federally funded Institute for Non-Lethal Defense Technologies (itself a part of Penn State's Applied Research Laboratory) are also pushing into pharmacological means of crowd control, listed among the 1992 U.S. Army's desired "disablements." They have identified a number of drugs that could incapacitate large numbers of people. The most promising drugs they have tested included "benzodiazepines like Valium, serotonin-reuptake inhibitors like Prozac, and opiate derivatives like morphine, fentanyl, and carfentanyl" (Arike, 2010, 47). Researchers acknowledged problems in moderating dosage delivery, but these might be solved, they suggested, through shared expertise of pharmaceutical industry partners.[20]

5.3 THE INCREASING DEPERSONALIZATION OF VIOLENCE

At risk of some simplification, we believe that one of the central historical thrusts of military technological development has been the extension of the distance across which kinetic force can be projected: the armored soldier, astride a horse, could project force over distance and defeat an adversary on foot, while the development of the long bow in the late medieval England would come to challenge armored horsemen. So too with firearms, the evolution of rifled barrels, battleships, and in the First World War, airplanes.

In 1991, social critic Jean Baudrillard extended this line of thinking in a series of essays entitled *The Gulf War Did Not Take Place*. While there is much to disagree with in his essays, one

[19] Amnesty International recorded 334 Taser-related deaths in the U.S. between June 2001 and September 2008. Of those killed, only 33 were armed (Arike, 2010, 46, note 15).

[20] Freedom of Information Act requests to further investigate these findings have been denied on national security grounds (Arike, 2010, 47).

argument stands out as particularly useful: that technology used by the U.S. and Great Britain in the Gulf War, especially "smart bombs" with cameras planted in the nose-cones, gave to those who watched that war on television news programs a sense of this as a "virtual" war rather than an actual war.

Such technologies allowed military operations to appear to North Americans and Europeans as little more than a television program, or perhaps a video game, and warmakers achieved a considerable domestic political victory: on television, war seemed to cost very little in terms of human life and to inflict only minor suffering. The bombs themselves and their purported precision became a media-driven marvel, while the enemy (to say nothing of potential noncombatants) receded to an afterthought. For the North American and European citizens of these liberal democracies, war became a troublingly sanitized affair, according to Baudriallard, "purged of any carnal contamination or warrior's passion. A clean war that ends up as an oil slick" (Baudrillard, 1995, 43).

The introduction of drone warfare in the Global War on Terror has made war even more "virtual" (from U.S. perspectives), even as media access has been radically restricted through "embedded" reporters. By no means do we diminish the real horror that soldiers on all sides of this now decade-plus-long conflict continue to experience or the ruined lives of non-combatants killed in drone strikes. These remain as real as they were when the idea of trauma was developed after the horrible intimacies of the First World War.

It is too soon to tell what sort of traumas drone operators, whose wartime lives are spent sitting in air-conditioned facilities and commute to sleep at home with their families, may confront when they come to reflect on the "bug splats," as operators have come to call drone-killings in Pakistan, Afghanistan, and elsewhere.[21] Clearly, the distance between political communities in whose name war is waged—and the places where actual battles rage—has become great. It poses profound ethical problems when killings are understood as nothing more than a smear across a video screen in a climate-controlled room.

5.4   DARPA'S SPIRAL OF INNOVATION

DARPA remains at the forefront of technology development in war-fighting, and it is likely the preeminent agency in recognizing, and perhaps, in the process, producing, new terrains of combat. DARPA's research is couched in the language of saving lives. As part of its mission, the DARPA's webpage proclaims, "The modern warfighter, whether on land, air or sea, must constantly innovate to keep one step ahead of the changing twenty-first century battlefield. Innovative research projects...are providing cutting-edge technology to U.S. military personnel and ultimately saving lives" (DARPA, 2012a). Perhaps the language of military innovation to save rather than take lives is

---

[21] On the language of "bug splats" to dehumanize those killed by drones, see Sara Waheed (2012).

unsurprising, given that DARPA is funded by an institution known as the Department of *Defense*, rather than the Department of *War*, a change in nomenclature made in 1949.[22]

DARPA's current promotional literature reads like something from science fiction—though it is hard to tell if it is a celebratory or dystopian future we peer into. There is a sense of wonder at the means of technological innovation at our (that is, American military scientists') fingertips. But there is also a sense of the profound danger facing the U.S. From cyber-security and information security to biological and robotic warfare and weapons sighting systems—the range of fields is astounding. But they share a common mandate based, curiously, around the idea of surprise: "[T]o prevent strategic surprise from negatively impacting U.S. national security and create strategic surprise for U.S. adversaries by maintaining the technological superiority of the U.S. military" (DARPA, 2012b).

5.4.1 MILITARY TECHNOLOGY PROLIFERATION

A bitter irony to the quest for surprise is that the endless proliferation of new military technologies continues to produce its own nightmares. Much as the Cold War nuclear arms race virtually guaranteed ever-escalating and excessive weapons systems that would more efficiently destroy the world—and indeed, led to the near-misses of accidental detonations (Scholsser, 2013)—DARPA's engineers and planners seem engaged in a race against themselves and the technologies they create. In his February 2012, testimony before the U.S. Congress, DARPA Deputy Director Kaigham J. Gabriel stressed looming dangers instead of his agency's successes. Though this strategy would help make the case for expanded research budgets, it also showed how military research has unpredictable geopolitical consequences.

Indeed, it appeared that DARPA had produced some of its own terrors, and the sources of new national vulnerabilities. The problem, Gabriel testified, was that many of the technologies DARPA helped develop were now widely accessible. "Computing, imaging and communications capabilities that, as recently as 15 years ago, were the exclusive domain of military systems, are now in the hands of hundred [sic] of millions of people around the world," Gabriel explained (DARPA, 2012c). While he lauded the benefits of these technologies—much of which had foundations in previous DARPA-funded research—he also stressed that nearly a dozen countries now use these store-bought technologies as means of electronic warfare. According to the DARPA webpage, "the pace at which these systems, formerly the purview of a few peer adversaries, are being developed is increasing."

The dreams of military planners are just as likely to manifest as their own nightmares, producing a spiral of spending, research, and programs with no end in sight.

---

[22] The reconfiguration that took place along with the name change also linked the different components of the US armed forces into a more coherently governable bureaucracy (Chambers, 1999, 167).

5.4.2   CONTEMPORARY MILITARY RESEARCH FUNDING

If these issues keep DARPA's planners up at night, they need not lose sleep over where their funding will come from. Indeed, money for their research is secure and growing. DARPA's funding in fiscal year 2001 was $1.98 billion. In 2002, it climbed to $2.25 billion, and in fiscal year 2003, it climbed again to $2.69 billion. Its funding for fiscal year 2013 is recorded at more than $2.8 billion dollars (DARPA, 2002; DARPA, 2012d). Much of this money is spent on far-reaching programs, with applications that only play out far into the future. One shudders to think of how research into the biomechanics and aerodynamics of flying snakes might be weaponized (Petrillo, 2011).

The most recent era has revealed funding structures similar to those in the Cold War, and defense spending continues to have a considerable impact on research priorities (Lucena, 2005). That the Department of Homeland Security has

> a $70 million dollar scholarship and research budget, and its initiatives, in alliance with those of the military and intelligence agencies, point toward a whole new network of campus-related programs. [For instance,] the University of Southern California has created the first 'Homeland Security Center of Excellence' with a $12 million grant that brought multi-disciplinary experts from UC Berkeley, NYU, and the University of Wisconsin-Madison. Texas A&M and the University of Minnesota won $33 million to build two new Centers of Excellence in agrosecurity.... The scale of networked private and public cooperation is indicated by the new National Academic Consortium for Homeland Security, led by Ohio State University, which links more than 200 universities and colleges (Martin, 2005, 27, as quoted in Giroux, 2007, 22).

This persistence suggests an ongoing need for critical engagement with those funding priorities, in the shaping of intellectual inquiry, academic research, and the goals of democratic states. The questions of warfare and the structural constraints it sets on inquiry, through either funding structures or tacit encouragement or dissuasion, are very much ongoing.

The two world wars and a global cold war among European powers unmade the 19th-century colonial world system. New technologies of travel allowed formerly colonized peoples to move to European metropoles in search of better economic lives. At the same time, military technologies shifted.

In the late-20th century, weapons planners determined that they would need to fight wars that minimized deaths. Their goals were less the acquisition of new territories, now concentrating on the extraction of economic resources. Kinetic weapons systems would continue to become more sophisticated and powerful, but they would be augmented by something of a post-Newtonian system of control. This system of control did not seek to destroy bodies through kinetic weapons, but instead to ensure the compliance of those bodies. This new emphasis gave rise to a more complex reckoning with human anatomies, and with the natural as well as social sciences. The ethics of engineering, of engineering education, and of the military-industrial-academic complex, have become vastly more complicated and, hence, must be closely scrutinized. Where each of us fits in this complex and shifting world and what our responsibilities and opportunities may be are open and evolving questions. But, as with the challenges surrounding integrity, these questions must be continually asked and re-asked. They must also be answered, however tentatively and imperfectly, by each generation of engineers, by engineering educators, and by engineering scholars. Some such answers—by would-be engineering reformers, both past and present—are provided in the following chapter.

CHAPTER 6

Responding to Militarism in Engineering

The prior two chapters have highlighted how the influence of militarism on engineering has remained strong even as historical circumstances have changed and as the end-goals of military technology research and development have evolved. This chapter returns to questions of ethical engagement of engineers around questions of militarism, especially as reform-minded engineers seek to transform the field away from one so heavily influenced by military interests—both by countering those influences directly or by nurturing alternative practices aimed at peace building.

Rather than attempting to catalog the many and diverse engineering peace-builders who might serve as models for rethinking engineering practice or guiding young engineers' career choices (see, e.g., Vesilind, 2005), this chapter has a different focus. This chapter directs attention to reform initiatives that seek to map and contest engineering's conceptual and structural frameworks that are amenable to militarism, particularly in response to the unique challenges imposed by modern technologies of warfare. To do this, we first identify two historical precedents of engineers seeking to redirect the field away from warfare and toward more humanitarian practices. Next, we look at some of the particular ethical challenges arising in the contemporary era given current trends in military technology development. Finally, we review contemporary responses by engineering reformers seeking to direct the field away from warfare in light of these particular ethical challenges as well as the enduring entwinements documented across this book.

6.1 HISTORIC RESPONSES: ANTI-WAR ENGINEERS

As with most occupational fields, engineering has always been constituted by members with diverse ideological orientations and social and political commitments. This is true even as the field as a whole has been characterized as relatively conservative and even though many engineers adopt an identity of purported political neutrality. Of those engineers who explicitly embrace particular social or political commitments, many have sought to reform engineering along the lines of those commitments.

Such explicit reform efforts within engineering have a rich history and have included efforts to steer engineering practice away from warfare and toward humanitarian ends. Particularly following the 1960s counterculture revolution, these efforts became visible, as engineers "looked for ways to make their political commitments fit more clearly with their professional lives" (Moore, 2008, 212). The radical politics of the time provided a notable force in stimulating debate within the sciences and engineering over professional responsibilities—particularly their responsibilities to the public good that existed beyond the immediate interests of corporate or government employers. At this time, "A confluence of circumstances...generated alternatives to accepted conceptions of technology—a moment of new possibility that momentarily appeared to alter what it would mean to be an engineer," where political commitments came to the center and widespread progressive practice seemed within reach (Wisnioski, 2009, 779).

Sometimes, reform efforts emphasized direct resistance to the influence of militarism on science and engineering. At other times, reformers emphasized preferred alternatives: humanitarian applications of engineering, particularly in response to the needs of marginalized social groups including the global poor. Both such approaches were evident during the politically tumultuous period of the late 1960s and early 1970s.

6.1.1   ANTI-MILITARISM REFORM EFFORTS

One important strategy for reforming engineering, connected to the 1970s counterculture movement, was direct action confronting the influence of militarism on engineering, including pointing professional and public attention to the growing influence of the military-industrial complex. This approach was taken by a group of leftist science and engineering reformers in the U.S. that created a series of overlapping organizations starting in the late 1960s. Their activities eventually crystalized into the Science for the People movement and, for a brief period, the Committee on Social Responsibility in Engineering (CSRE). As with the larger counterculture movement, resistance to the Vietnam War was a primary organizing impetus for both of these groups; however, more generally, they sought to rectify what they considered to be the widespread *misapplication* of science and technology.

Scientists and Engineers for Social and Political Action/Science for the People

An early step in what was to become the Science for the People movement was the formation of the group Scientists for Social and Political Action (SSPA), which held its first gathering at the 1969 American Physical Society meeting in New York. In the Call for Participation (SSPA, 1969a) for this gathering, the founders of the group stated: "we now see that many of the products of science and technology have become more a menace than a boon to the interests of human society" and called for the creation of "an independent body of socially aware scientists free from the inhibitions which abound in the established institutions we now serve." This was not a call for increased attention to professional responsibility in an abstract sense, but a call to political *action* in response to the professional associations' inclination to remain "aloof from the desperate problems facing mankind today."

Among the most pressing of the problems facing humanity, according to the group, was the inordinate role of military interests in directing science and technology research as well as the lack of concern over this situation by professional associations. In the group's first newsletter following its 1969 formation at the American Physical Society meeting, three broad problems were identified as guiding the group's work: 1) militarism, 2) technologies contributing to environmental and social problems, and 3) lack of organizational leadership within the sciences to respond to these problems. Concerning the problem of militarism specifically, the group stated: "Government support of research and development is overwhelmingly dominated by military projects, while the existing levels of armament already constitute the greatest threat to world peace and security. The ABM [anti-ballistic missile, the major new weapon system of the day] program must be stopped" (SPPA, 1969b, emphasis in original).

Later in 1969, the group changed its name to include engineers explicitly, becoming Scientists and Engineers for Social and Political Action (SESPA), and still later, at the December 1969 meeting of the American Association for the Advancement of Science, SESPA members coined the term "Science for the People" to describe the group's overarching goal. "Science for the People" also became the name of SESPA's newsletter-turned-magazine starting in 1970 and continuing through 1989 (Schwartz, 2011). At its peak in 1974, *Science for the People* (the magazine) had over 4,000 subscribers (Moore, 2008).

Besides *Science for the People*, one of SESPA's major publications was the 1972 report, "Science against the People," an exposé of the Jason program, documenting the program's history and structure as well as profiling some of its affiliated scientists. The Jason program was an effort by the U.S. Defense Department to attract top academic scientists to carry out advanced military research. The program was run by a newly created, quasi-independent research institute, the Institute for Defense Analysis, which contracted with academic scientists in a way that gave them considerable administrative freedom in their work while aligning it with military research goals. "Science against the People" brought the initiative to light among scientific communities and the wider public, but received considerable criticism for its strategy of publicly shaming Jason scientists and generally portraying them in a disparaging way (Moore, 2008).

Of particular interest to our analysis, however, is that "Science against the People" explicitly addressed the *political* dimensions of scientific military research in contradistinction to participating scientists' rhetoric of neutrality. "One key theme that [SESPA] emphasized was that the Jason scientists were not politically disinterested, but allowed their political views to shape their decisions about weapons" (Moore, 2008, 171). Further, as "Science against the People" argued, military research scientists were not "simply following orders," but often played a leadership role in advancing new weapons systems research:

> [I]t should not be thought that these scientists work only at the instigation of the military; quite the contrary, the most novel weapons can not be anticipated by non-scientists and are often resisted by a conservative majority of career soldiers. The atom bomb, the hydrogen bomb, intercontinental missiles, nuclear submarines, chemical and biological agents, the

automated battlefield—all of these had, and needed, first-rate scientists to champion them, not just to supply them to the Pentagon's order. (SESPA, 1972, Chapter 3)

Perhaps reflecting waning interest in the *organizational reform* of science and engineering professional associations—despite continued interest in *politically directed* science and engineering per se—*Science for the People* removed its reference to SESPA on the front cover in 1975 and then, in 1977, dropped the SESPA membership form from the rear cover, replacing it with a *Science for the People* subscription form instead. As a result, one's subscription to the publication was divorced from organizational membership in SESPA. The goal of reforming science and engineering *professional associations* that was originally at the heart of SESPA seems to have died away as supporters' attention shifted to providing a public venue for alternative visions of science more generally.

But this latter achievement—providing a public versus professional venue for science criticism—turned out to be a great success of SESPA. As historian Kelly Moore points out, the Science for the People (SftP) movement "made it difficult to treat scientific knowledge as distinct from the power relations that produced it…. SftP [showed] that the values and beliefs of scientists, their sponsors, and those who use science ought to be included in debates about the veracity and social value of scientific claims" (Moore, 2008, 187). From their early focus against the Vietnam War to their two-plus decades of publishing *Science for the People*, SESPA played a critical role exposing the institutional structures that direct scientific work and the need for individual, professional, and public reflection and deliberation around the extent to which such shaping is appropriate.

Committee for Social Responsibility in Engineering

At the same time as SESPA was getting its feet, an overlapping group of engineers committed to similar goals focused its attention specifically on the domain of engineering. This group formed the Committee for Social Responsibility in Engineering (CSRE), which published its own magazine, *SPARK*, from 1971 to 1974. CSRE's politics and strategies were closely aligned with SESPA's and were directed largely at the growing military-industrial complex, but CSRE also included considerable attention to issues surrounding engineering employment and labor issues (Wisnioski, 2012). In their "Statement of Purpose," published in the first issue of *SPARK*, the editors put forward their position on militarism in stark terms: "Thousands of engineers feel that their engineering talents are misused in both civilian and military projects, and believe that the constant development of weapons technology spells the ultimate disaster of mankind" (CSRE, 1971, 3).

Through *SPARK*, CSRE members grappled with many of the tensions surrounding their leftist political commitments and the structural reality of an engineering profession influenced considerably by corporate capitalist and military interests. They pointed to the moral dilemma resulting from the fact that science and technology served as a foundation for militarism worldwide (CSRE, 1971, 18), and how some specific members responded to the conflict of working on military projects (CSRE, 1973, 2). As with SESPA, they also exposed specific structural dimensions of the military-industrial complex as it played out in engineering, for instance by noting the pervasive overlap in directorship of government agencies and defense-industry corporations, notably in the Department of Defense and corporations such as General Electric, Hewitt Packard, Honeywell, and International Telephone and Telegraph.

CSRE unraveled in the mid-1970s, just as SESPA began its transition into the more general Science for the People movement. In addition to the dissipation of the counterculture movement's momentum and the winding down of U.S. investment in the Vietnam War, CSRE members' political commitment to decentralized decision making and the lack of a focused organizational strategy and a clear leadership structure contributed to these changes (Moore, 2008). Some members of CSRE redirected their energies to the Committee on Social Implications of Technology (Unger, 2013)—a committee of the Institute of Electrical and Electronics Engineers (IEEE, the world's largest engineering professional association)—which persists today under the banner, Society on Social Implications of Technology.

Unlike SESPA's *Science for the People*, CSRE's reform efforts were and remain less widely known—focused as they were more directly on professional reform inside engineering. But like SESPA, CSRE provided both precedent and conceptual insight around the inherent politicization of all engineering work and the existence of often-divergent notions of the public good circulating within the profession.

6.1.2   HUMANITARIAN REFORM EFFORTS

A very different reform strategy was taken by advocates of the appropriate technology movement, which was perhaps one of the more successful efforts to redirect engineering activity away from the military-industrial-academic complex and toward "humanitarian" goals during the 1970s. This movement, spanning the 1970s and early 1980s but with roots going back further, was far less focused than CSRE in its reform agenda, emphasizing alternative technologies generally and supporting *alternative technical practices* over direct political confrontation within established political and professional institutions. It sought to redirect technologies toward a broad vision of the public good specifically in the context of economic and technological inequities (Willoughby, 1990).

The appropriate technology movement offered a critique of and corrective to dominant economic development models of the time, centered as they were on Western-style consumerist industrialization and a simplistic model of technology transfer. On one hand, the existing model could be seen as fitting into forms of economic development championed by U.S.-based Cold War theorists like Rostow as a kind of anti-Soviet political and ideological strategy. On the other, this simplistic model literally and figuratively dropped advanced technologies developed in rich countries into poor contexts, expecting them to function equivalently. In response, appropriate technologists were early in recognizing that different contexts had different opportunities and constraints, requiring different technologies and supporting social practices (Dunn, 1978).

In the seminal book *Small is Beautiful* (1973), E. F. Schumacher popularized the concept of appropriate technology by promoting an alternative to then-current high-tech versus low-tech options for pursuing economic development in poor economies across the globe. For Schumacher, technologies that were "intermediate" between low and high tech were more appropriate in most global contexts of poverty—more appropriate to the needs and means existing within those contexts. Specifically in the Asian contexts that Schumacher studied, this usually entailed inverting the high-capital, low-labor ratio guiding technology development in rich countries. But his approach was more nuanced than this. As Schumacher points out:

> It is too often assumed that the achievement of western science, pure and applied, lies mainly in the apparatus and machinery that have been developed from it, and that a rejection of the apparatus and machinery would be tantamount to a rejection of science. This is an excessively superficial view. The real achievement lies in the accumulation of precise knowledge, and this knowledge can be applied in a great variety of ways, of which the current application in modern industry is only one. The development of an intermediate technology, therefore, means a genuine forward movement into new territory, where the enormous cost and complication of production methods for the sake of labour saving and job elimination is avoided and technology is made appropriate for labour-surplus societies. (1973, 198)

The practice of appropriate technology expanded greatly in the years around the publication of *Small is Beautiful*. In the U.S., the group Volunteers in Technical Assistance (VITA) was formed in 1959 to provide international technical development assistance, directing attention to what became known as appropriate technologies in the 1970s. Similarly, in the United Kingdom, Schumacher collaborated with Georgie McRobie and Julia Porter to found the Intermediate Technology Development Group in London in 1966. After over five decades, both of these organizations continue their work, albeit in modified configurations.[23]

The literature that supported and came out of the appropriate technology movement included theoretical work focusing on how and why appropriate technology offers a genuine alternative to dominant technological practices. As with CSRE and SESPA, appropriate technologists recognized that technologies have embedded power relations, and hence different technologies can

---

[23] VITA merged first with EnterpriseWorks (in 2005) and then with Relief International (in 2009), which focuses on disaster relief while considering long-term developmental needs (accessed May 31, 2013 at: http://www.enterpriseworks.org/display.cfm?id=2&sub=1). ITDG changed its name to Practical Action (in 2005), in part to emphasize the need for broad-based approaches to development assistance: "Technology has a vital role to play in building livelihoods—and technology can include physical infrastructure, machinery and equipment, but also knowledge, skills and the capacity to organise and use all of these" (accessed May 23, 2013 at: http://practicalaction.org/what-we-do-7).

enable different types of social relations (see, e.g., Dickson, 1974). Common in the late 1960s and early 1970s—when the limitations of the dominant technology transfer models were first becoming widely apparent and the social movement surrounding appropriate technology was gaining momentum—these works were strong on advocacy but weak on critical inquiry into the potential shortcomings of their vision of alternative practice.

After the 1970s, the appropriate technology movement faded considerably in the global North, but it thrived in the South, and the scholarship surrounding appropriate technology matured all the while. With historical perspective, careful theoretical and historical analyses of appropriate technology as a concept and as a social movement became more prevalent (e.g., Smillie, 2000; Willoughby, 1990). Taken as a whole, the framework of appropriate technology offers a compelling contribution to theoretical considerations of the politics of technology and technical expertise. It also provides precedent for engineers seeking to direct their efforts toward community development work and shines light on the challenges such work confronts as well as the wide range of skills needed to effectively respond.

One particularly important lesson from the appropriate technology movement is that specialized technical knowledge is often used to overrun local knowledge systems and priorities, even when intended to do the exact opposite (Nieusma and Riley, 2010). Hence, the best of appropriate technology practice negotiates this tension—explicitly and robustly—in working toward community empowerment. Unfortunately, while the appropriate technology movement builds on an explicit political agenda around economic equality and social empowerment, and while it became popular alongside other counterculture movements of the 1970s and 1980s, much of the work on appropriate technology during this time failed to grapple with the macro-political context of development work. Hence, appropriate technology as a recognized social and political *movement* faded considerably in the 1980s, despite the continued success of numerous appropriate technology initiatives and organizations.

Like CSRE and SESPA, enduring lessons were learned through the efforts of appropriate technologists. Perhaps the most important, if general, of these lessons involves the *centrality of its particular social context* in assessing a technology's appropriateness for achieving social (including economic) goals. In practice, this meant engineers working on development projects needed to learn "on the ground" how their technologies fit within varying contexts. In theory, the lesson entailed rejecting universalist narratives of progress, in terms of both technology development and national economic development; there was no single, linear path to development that all nations must follow. The appropriate technology movement's concept of "technology choice"—that the most appropriate technology to a given social or economic development problem is neither universal nor self-evident—provides historical backdrop to much of the work around sustainability and sustainable technology design today.

As the 1970s engineering reform efforts reviewed above show, "Engineers are not preordained to reproduce the status quo, and a particular technological future is not inevitable" (Wisnioski 2009, 413). But challenging the deeply embedded structures that align the interests and worldviews of engineers, large corporations, nation-states, and their militaries is a nontrivial endeavor. As we noted in Chapter 1, questions surrounding the ethical implications of militarism and war on engineering are infrequently discussed—surprisingly so—within engineering education and engineering professional associations today. This is true despite the variety of efforts by engineering reformers in the 1970s and 1980s highlighting the political nature of engineering work. Obviously, enduring structures—conceptual and institutional—constrain change along this dimension.

Part of the challenge of addressing militarism in engineering in the contemporary context is *blurring of the boundaries*—conceptual, institutional, and technological—that have traditionally demarcated discrete domains of action and responsibility. The muddling of institutional boundaries, particularly around the military-industrial-academic complex was discussed in prior chapters. The remainder of this chapter will consider the blurring of technological and conceptual boundaries as related to militarism in engineering. The next section explores ethical challenges arising specifically in response to the blurring of technological boundaries as the nature of military and warfare technology research changes. Then, the chapter's final section explores contemporary reform efforts in engineering that seek to tackle the conceptual blurring in engineering that confounds change within the profession.

6.2   ETHICAL CHALLENGES OF CONTEMPORARY WARFARE

While the engineering reform movements of the 1970s and 1980s grappled with many of the challenges surrounding warfare technology development and the military-industrial-academic complex, they did not anticipate the changing cultural landscape that would eclipse their efforts to expose engineering work and its output as always politicized. To be sure, some critics of militarism within engineering remained vocal, and groups such as IEEE's Society on Social Implications of Technology have continued to interrogate the of militarism on engineering, but the radical engineering reform movements of the late 1960s and early 1970s waned in the late 1970s and early 1980s. During this time and since, of course, warfare technologies have continued to evolve.

For the most part, the development and application of contemporary warfare technologies can be viewed as motivated by strategic rather than ethical concerns. The science-fiction-like character of technologies and weapons associated with the future of war—robotics, drones, armor, and advanced chemical and biological compliance technologies—are directed as much toward the curtailment of military casualties on one's own side as they are toward the destruction of the enemy. One realistic account of the reason for this trend is that any conflict resulting in heavy casualties is likely to be politically unsustainable over time.

... at such technology saves lives and limits destructiveness is an illusion masking
... nd nature of the destruction.

... he illusion of limited destructiveness are other, related and even greater dangers.
... soldiers from war may be seen as the logical limit of what military planners, and
... rs who develop military technologies, have tried to do from the earliest times. With
... ever-more-powerful weapons, delivered with greater accuracy from greater distances,
... elf has been carefully engineered to move away from hand-to-hand and face-to-face
... this way, military engineering is responsible for removing the enemy from one's direct
... is easier—for psychological among other reasons—to kill from a distance, because it is
... this way to imagine and construct an idea of those against whom one is fighting as "wholly
... and different from oneself.

... Erich Maria Remarque's novel about the horrors of World War I, *All Quiet on the Western
... nt* (1929), is famous for making this exact point but in a way that humanized the enemy. It is a
... enario often copied in literature, film, and television by way of didactically illustrating how deep-
... eated hatreds and prejudices might be overcome via direct confrontation. In facing one's enemy
... directly, one may come to see that the differences between oneself and the "other" are basically illu-
... sory and manufactured—a product in part of what we are taught. Paul Bäumer, a German soldier,
... is talking to Gérard Duval, the French soldier he has just killed and who lies before him.

Comrade, I did not want to kill you.... But you were only an idea to me before, an ab-
straction that lived in my mind and called forth its appropriate response.... I thought
of your hand-grenades, of your bayonet, of your rifle; now I see your wife and your face
and our fellowship. Forgive me, comrade. We always see it too late. Why do they never
tell us that you are poor devils like us, that your mothers are just as anxious as ours, and
that we have the same fear of death, and the same dying and the same agony—Forgive
me, comrade; how could you be my enemy? [223]

Military technology has endeavored, with partial success, to prevent others from seeing what
Bäumer is able to see in virtue of the proximity of the person he has just killed. It has done this by
enabling killing at safe distance, that is a distance where the illusion of an "other" wholly different
from oneself can endure. The reason for this—and the moral of the incident as Remarque tells
it—is clear. Geopolitics had taught, and distance had enabled, him to hate his enemy. But havin...
seen what he saw and having had that conversation with the man he just killed, Bäumer could...
continue to kill his "enemies," who turn out not to be all that different from himself. Killin...
distance precludes this eventuality.

6.2.2 THE BLURRING OF MILITARY TARGETS

In terms of killing at a distance, the advent first of long-range artillery and later of w...
military engineering milestones, making it not merely possible but routine to con...

As a matter of political reality, a war cannot result i
deemed unacceptable to the public. If robots and dr
then the result is an expense rather than a casua¹
ment, but a profit for the firm that manufactur
a casualty. Weapons technologies that can kill wit.
give politicians leeway and, hence, more options for ε
not otherwise be politically sustainable.

It is perhaps ironic that to the extent that soldiers
robotic, cybernetic, and other high-tech devices, it is engineers
without having to be physically located at or even near the site
new frontline agents of warfare. It is engineers and engineering te
drones, instructing the robots, developing weapons guidance and de.
Hence, in an all-cybernetic war, the opposing armies are not artificial inte.
drones, etc., though it is made to look like that in science-fiction films. Inst
and technicians who invent, perfect, maintain, and operate them. While enginee.
behind the development of technologies of warfare, as was highlighted in the first
of the very formation of the concept of engineer, newer technologies remove more an
"middlemen" in warfare. Engineers are increasingly becoming direct participants in acts

In this respect, engineers are displacing traditional soldiers, and at a rapidly increasi.
If an all-cybernetic and robotic military were available today, it is clear that those in power w
choose such an "army" wherever feasible. And in so choosing, they would actually be enlisting e.
gineers as frontline soldiers.

6.2.1 THE HIDDEN VIOLENCE OF DEPERSONALIZED WAR

Contemporary engineering's complicity with efforts to depersonalize war raises particular ethical
challenges. It may seem that this depersonalization of war—wars where ideally no one on one's own
side is killed, and no one on the other side is "seen"—is wholly desirable. It is obviously better to
lose a drone than a pilot and crew. The problem of this approach is—as alluded to above and as any
viewer of science fiction movies involving war and violence knows—the use of such devices may not
only fail to limit the destructiveness of war but can help increase it. Causalities on one's own side
may be temporarily curtailed, but the increased destructive capacity of such weaponry and technology, along with the fact that the opposing side will also have such technology to some extent,
facilitates an even greater destructiveness than would be possible with a more traditional military.

Again, insofar as such technology gives the illusion that fewer people are being killed than
might otherwise be, it exacerbates destructiveness by increasing the military options available to
those waging war—for example by making it politically possible for them to conduct a lengthier

78 6. RESPONDING
war. The notion th
both the degree
Beyond t
The removal o
hence engine
inventions o
warfare its
combat. I
sight. It
easier t
other"

Fro
sc
s

out ever having to see who it is one has injured or killed or otherwise experience the destructive force inflicted. Drones and similar weapons take the illusion of non-destructiveness a step further. Particularly drones, but also surveillance technologies generally, make it possible to conduct a very different kind of warfare, for example, by seeking out and targeting specific individuals to be killed. Such capability enables states to assassinate individuals, including non-combatants, who have not been convicted of any crime. Given that new technologies usually proliferate, and sometimes rather quickly, it is reasonable to expect that many countries and, soon, individuals will be using similar technologies to assassinate those whom, for whatever reason, they deem necessary or desirable to eliminate.

There has also been a blurring of the lines between military and civilian security and law-enforcement agencies, which has its own ethical, legal, and social implications. Drone technology and pervasive surveillance systems have made their way to civilian police forces. Particularly since 2001 and the introduction of the U.S. "Patriot Act," civil-liberties organizations have warned about the dangers of unwarranted invasion of privacy and abrogation of civil and human rights. Such concerns have come to a head with the 2013 release of sensitive U.S. surveillance policy documents by National Security Agency whistleblower Edward Snowden. Whether or not one thinks increased governmental securitization is a good thing, there is no doubt that widespread monitoring has been made easily possible, largely an outcome of prior military research.

In successfully and increasingly distancing those who are killing from those killed in conflict, military and otherwise, engineers who have developed such systems have, however inadvertently, increased their own moral responsibilities. Indeed, given the roles they have assumed, there is no hyperbole in claiming that some engineers might accurately be seen as non-uniformed, frontline combatants. In this regard, advanced technologies demand a rethinking of warfare itself. We reiterate that killing and kinetic violence have not diminished in warfare or political suppression. But, in addition, in this new mode, the "frontline" must be understood as a connective interface rather than a geographic barrier, a fiber-optic cable that connects instead of a fortified line that divides.

The trajectories taken by modern military technologies—in particular the shift to depersonalized violence and non-lethal compliance techniques—create new complexities for engineers who wish to identify career paths with integrity. These trajectories show that one need not fight on the frontlines to participate in warfare and, more subtly, that one need not "build bombs or guns" to participate in the development of warfare technologies. "Soft-kill" is still a form of violence and control. While there may be moral benefit in moving away from outright killing, this benefit is achieved with a trade-off. Soft-kill enables the extension of military technologies to broader populations, including not just non-combatant participants in war, but also non-compliant populations of any sort engaged in civil disobedience.

6.3   CONTEMPORARY RESPONSES TO ENGINEERING AND WAR

This section explores a few modern-day engineering reform initiatives that seek to redirect engineering away from warfare, militarism, and technologies of control and suppression. It looks specifically at contemporary analogues to the 1970s movements discussed above: What we call "Engineering for Peace and Justice" initiatives have strong parallels with the Committee for Social Responsibility in Engineering and the Science for the People movement, and the more widely referenced "Engineering for Sustainable Community Development" initiatives parallel work of the earlier appropriate technology movement. A variety of efforts can be categorized under these contemporary initiatives, only some of which will be reviewed.

Beyond describing these contemporary reform initiatives in engineering, we will also assess the ways in which such efforts extend, and can be distinguished from, the prior movements. We will consider the ways they respond to militarism and the evolving nature of military technology development as well as the particular challenges arising from the military-industrial-academic complex as it stands today. Most importantly, besides providing alternative, non-militaristic visions of engineering practice, these reforms efforts seek to confront both the structural conditions and the conceptual boundaries around engineering that constrain reform initiatives. Practicing engineering with integrity, we will go on to argue, requires engineers to engage such questions, both individually and as a profession.

6.3.1   ENGINEERING FOR PEACE AND JUSTICE

Numerous initiatives in engineering reform over the past couple of decades have directly tackled the influence of militarism on the field and sought to promote peace and/or justice as alternative conceptual frameworks. Even more so than SESPA and CSRE before it, the Engineering for Peace and Justice "movement" is less a coherent, coordinated professional reform initiative and more an on-going thread of concern, inquiry, and advocacy within the engineering profession and engineering education circles. Over the last decade, two strands of this thread stand out: P. Aarne Vesilind's and W. Richard Bowen's initiative around "peace engineering" and the wide-ranging work of members of the Engineering, Social Justice, and Peace network.

Peace Engineering

Civil engineer-turned-engineering ethicist P. Aarne Vesilind opens the preface to his edited book, *Peace Engineering* (2005), with the following concern:

> Engineering has historically provided the tools necessary for defending our country and for promoting freedom throughout the world. But do engineers also have a central role in the promotion of peace—a peace that might make such war unnecessary? That is, can

engineers use their education and skills to proactively encourage peace in our world—to practice peace engineering? (xi)

Interest in the possibilities for and practice of "peace engineering," among Vesilind and his engineering colleagues at Bucknell University, led to a 2003 symposium titled, "Engineers Working for Peace," which laid the foundations for Vesilind's book and its later revision with Bowen. *Peace Engineering* covers a range of topics at the intersection of engineering and militarism, including warfare and weapons research, terrorism, environmental justice concerns, engineering for international development, and engineering education and university life. Ultimately, Vesilind's goal with the work is to identify both the wide range of skills—"technical...social, political, communication, ethical, and legal" (xi)—and the opportunity structures needed to practice the "new profession of peace engineering" (1).

In addition to identifying the conditions needed to practice peace engineering, Vesilind has also investigated the historical influence of militarism on the engineering profession in *Engineering Peace and Justice* (2010). This later work tracks the engineering profession from its military roots, through the birth and growth of "civilian [i.e., not military] engineering," and on to the present attention to engineers' environmental and professional responsibilities, wherein he sees the germination of peace engineering. *Engineering Peace and Justice* also identifies several "exemplars" of peace engineering, those whose work might provide productive models for young engineers seeking to practice this new profession.

Notably, Vesilind's exemplars include a dozen engineers working on environmental and international development projects addressing the subsistence, health, and safety needs of vulnerable populations around the globe. One additional exemplar represents what Vesilind calls an "apparent decision of conscience" (2010, 144) to reject military engineering research. Peter Hagelstein, who invented the technological basis for the Reagan administration's so-called "Star Wars" missile defense program (officially, the Strategic Defense Initiative), resigned from his post at the Lawrence Livermore National Laboratory to work instead at Massachusetts Institute of Technology "on research to 'benefit all mankind'" (2010, 144). Certainly, there are other contemporary engineers who have opted out of military research (see, e.g., Faure-Brac, 2012), even as they remain obscured in the professional ranks.

The work by Vesilind, Bowen, and their collaborators offers a sustained contribution to contemporary conversations around engineering for peace and justice, specifically by extending the logic of professional responsibility into the domain of "peace-building." The goal of this work is not to systematically explore the structural conditions and professional assumptions that align engineering education and engineering professional practice with military interests and outlooks, but to create imagination for an alternative set of practices and assumptions that are oriented explicitly toward peace. This work also provokes conversation about militarism and war in engineering ethics circles, which, as noted in Chapter 1, tend to eschew the topic.

Most provocatively, Vesilind proffers peace engineering as a new *profession*, not just one among many applications of engineering expertise. As such, it raises a series of important questions both for individual engineers navigating their own careers and for reformers seeking to re-envision the field as a whole. These two, related concerns are captured in the subtitles of Vesilind's aforementioned books—*When Personal Values and Engineering Careers Converge* (from *Peace Engineering*) and *The Responsibility of Engineers to Society* (from *Engineering Peace and Justice*).

Engineering, Social Justice, and Peace

Whereas peace engineering focuses on alternative practices that exist at the margins of the engineering profession, members of the Engineering, Social Justice, and Peace network tend to focus more on the dominant political, economic, and conceptual structures within engineering that keep alternative visions of engineering, such as Vesilind's, at the margins. In other words, while work under the banner Engineering, Social Justice, and Peace is diverse, defining elements of the approach is attention to 1) dominant, enduring power relations within engineering; 2) how those relations systematically constrain (and, hence, outlive) individual reform efforts; and 3) strategies for responding to the systemic constraints.

"Engineering, Social Justice, and Peace" (ESJP), as a moniker and an established network of reformers, extends from the work of materials engineer and engineering education expert Caroline Baillie, who hosted the "Engineering for Social Justice" conference in 2004 at Queen's University in Canada. Attending that event was aerospace engineer-turned-environmental ethicist George Catalano, who hosted a follow-up event titled "Engineering, Social Justice, and Peace" in 2006. Since the 2006 meeting, members of the ESJP network have taken turns hosting an annual conference, and the network itself has grown to include practicing engineers, engineering educators, and engineering studies scholars representing numerous engineering and related disciplines, such as ethics, communications, and technology studies.

ESJP exists as a loosely affiliated network, with a coordinating committee serving to organize activities and ensure continuity. In addition to the annual conference and wide-ranging scholarly collaborations among its members, the network publishes a peer-reviewed, open-access journal, the *International Journal of Engineering, Social Justice, and Peace*; produces the zine, *Reconstruct*; and offers additional resources on its website (esjp.org).[24] In addition to these outward-facing resources, ESJP strives to provide members and supporters a nurturing community and helpful environemnt in which to carry out reform-oriented research and practice, not least including in engineering education.

---

[24] Given the relatively small sizes of the engineering-for-peace-and-justice communities, it is not surprising there is considerable overlap in their networks and across their collaborations: This book is the direct outcome of one such collaboration; co-author Nieusma serves on the ESJP coordinating committee, as do Baillie, Catalano, and several others cited throughout this book; and some of the approaches to engineering reform reviewed here and throughout are spearheaded by various members of the overlapping communities.

The coordinating committee of ESJP created a statement of commitments to help orient visitors to the ESJP website as well as potential participants in ESJP activities. The statement includes a number of political and intellectual commitments that serve to orient the network, even as there exists divergence within the community. One such commitment stands out in the context of this analysis:

> We are committed to working for peace and nonviolence—within ourselves, our group, the engineering profession, and the wider world. We understand peace and social justice to be mutually constituted, each requiring the other to be meaningful. We are unsettled by the close relationship between engineering and militarism, both historically and in the present, and are committed to engineering that alleviates suffering caused by violence of all sorts. (ESJP, 2010)

Beyond its commitment to peace and nonviolence, much of the scholarship produced by members of the ESJP network is relevant to our analysis, especially insofar as this work grapples with the *deep structuring forces* that align engineering thinking and practice with militarism.

The deep structuring forces that ESJP work highlights include both direct and indirect influences of militarism on engineering and vice versa. As we have pointed out in preceding chapters, patronage structures ensure a high percentage of engineering work is carried out to achieve military goals. Conversely, engineers contribute directly to the conceptualization, design, and development of technologies of warfare, surveillance, and control and they otherwise contribute to the successful running of organizations that are embedded in the military-industrial-academic complex. Notably, relatively little work by ESJP members covers this terrain systematically, with several notable exceptions identified here and in Chapter 1.

In addition to direct influences, however, are the indirect ways militarism shapes engineering, both as a field of practice and as an approach to solving problems. As Nieusma and Blue (2012) have shown in a prior collaboration, militarism manifests itself in engineering thinking and practice in subtle ways that extend beyond direct contributions to weapons research and development. As described by Noble (1979) and Streett (1993) and discussed in the introduction, the physical and mental disciplining of engineering students, programmed into engineering curricula throughout the mid-20th century, was directly and explicitly taken from military training, and it remains to this day a subtle but pervasive cultural barrier to engineering educational reform. Derived from this history is the challenge, as Riley (2008, 42) points out, of engineers' "uncritical acceptance of authority," again derived from military training and its strict adherence to chain-of-command decision making.

Although only a limited amount of ESJP scholarship addresses engineering and warfare systematically, the approach to social justice and peace taken by most ESJP scholars can be applied directly to the challenges of militarism. Most generally, ESJP scholars tackle the "structural problems with the profession—its military and corporate focus and the narrowness of engineering education, which excludes a number of important skills—[that] can present obstacles when we engage in social justice work" (Riley, 2008, 43).

ESJP scholarship also directs attention to the conceptual constraints that resist social-justice inspired professional reform. As Lucena, Schneider, and Leydens put it, the goal is to "[reveal] the hidden assumptions" that engineers bring to their work (2010, 56), including their social-justice-oriented work. Similarly, this time in Baillie's words, engineers are trained for "compliance" and are not provided the critical analysis tools needed to question the field's disciplinary assumptions—its "common sense" view of the world (2006, 4). Across these approaches, we see "attention to the 'dominant worldview' surrounding engineering...and how it contributes to a sense of inevitability surrounding engineering practices and outcomes" (Nieusma, 2013, 35–36).

Work in the domain of "Engineering for Peace and Social Justice" highlights alternative practices to militarism and calls attention to the conceptual and structural conditions that resist social-justice-and-peace oriented reforms. While a narrow focus on militarism and warfare in engineering is not common in this work, militarism and the military-industrial-academic complex are consistent undercurrents—both in reform efforts and in the scholarship that results from and contributes to these reforms.

Combining attention, on one hand, to alternatives to militarism in engineering practice and, on the other hand, to the conceptual and structural barriers that impede the pursuit of such alternatives offers a unique opportunity for synthesizing critical analysis with personal reflection and one's own ethical commitments in weighing engineering career options. Whereas the 1970s engineering reform efforts advanced thinking about engineering by highlighting its in-built politics and some alternatives, engineering for peace and justice seeks to take the next step by re-envisioning the entire field of engineering in a way that routinely enables such "alternatives" to be realized.

6.3.2   ENGINEERING FOR SUSTAINABLE COMMUNITY DEVELOPMENT

In the 1980s and 90s, interest in appropriate technology gradually shifted toward interest in environmentalism, "green" technology, and sustainable development, especially among progressive engineers in the global North. This interest has proven to be both widely appealing and enduring, undoubtedly enabled by its relatively generic commitments and its professed universal applicability (think of the metaphor, "spaceship Earth").

An example of sustainability's generic and hence broad appeal is provided by the oft-cited definition of "sustainable development" provided by the 1987 Brundtland Report, *Our Common Future*: "development that meets the needs of the present without compromising the ability of future generations to meet their own needs" (Brundtland, 1987, Chapter 2, paragraph 1). Another widely accepted framework for sustainability is the "triple bottom line" approach, which measures organizational success according to financial, ecological, and social benefits. This framework is also called "the 3 Ps" approach, which refers to "people, planet, and profit."

Since these approaches to sustainability were first popularized, interest in sustainability has continued to grow—as measured by number of advocates, their approaches to achieving sustainability, and the diversity of their underlying social and political commitments. The theme of sustainability has also begun to pervade academic programming, including in many engineering programs, despite what many education critics consider to be ossified disciplinary structures that quash change initiatives. Given its uptake and enduring presence, there is good reason to be enthusiastic about sustainability as an impetus for reform of engineering, including by shifting the field away from warfare and toward the broader public good.

However, there is also justification for caution in declaring the demise of "unsustainable" ways of living and thinking. The popularity of sustainability as a cultural theme has been so widespread largely because of its generic commitments—e.g., to future generations, or to "people" alongside profit; who would disagree with these?—and the extent to which these commitments can be coopted by consumer capitalism. Despite the ironic twist, sustainability *sells*. Cynicism is extended—and rightly so—when environment-ravaging corporations, and even entire industries (e.g., petroleum and coal), enthusiastically embrace "sustainability" with sophisticated, multi-million-dollar "green-washing" campaigns: BP's "beyond petroleum" catchphrase and the massive "clean coal" initiative.

These contradictions notwithstanding, attention to sustainability and sustainable development in engineering (and elsewhere) continues to offer promise for systemic reform. Pioneers in sustainability engineering provide productive examples of alternative practice, and wide-ranging analytic tools have been developed and integrated into engineering education and practice, from life-cycle analysis to biomimicry to product-service substitution. Rather than identifying the diversity of such tools, or even characterizing the influence of sustainability on engineering, our purpose in this section is more narrow: To highlight one promising thread in the domain of engineering for sustainable development, namely the turn toward "community" that this work has recently taken.

In their book on this exact theme, "Engineering and Sustainable Community Development" (2010), Lucena, Schneider, and Leydens suggest that attention to "community" is a critical reorientation of the sustainable development framework for engineers. As they put it, "*community needs to be central* to development projects, particularly, to the engineering dimensions of these if we hope for these projects to be sustainable in the long run and to increase the self-determination of the communities that they are intended to serve" (2010, 5, emphasis in original). The appropriate technology movement shifted attention from technology's productivity to its *appropriateness to context*—economic, environmental, and socio-cultural. Engineering for sustainable community development builds on this logic by systematically accommodating community members' perspectives and priorities as a fundamental component of the "local context."

Attention to *community* in sustainable development initiatives carried out by engineers provides a corrective to much of the engineering-for-development activity that has taken place since the appropriate technology movement's heyday. Innumerable projects in the spirit of "engineers without borders"[25] advanced what were nominally appropriate technologies, and implemented them in communities around the globe. Too often, however, this was done with little interaction with—and too little understanding of—host communities or their members.

By repeatedly and systematically failing to understand community members' perspectives and priorities around "development," and by imposing a foreign notion of "the community's needs" in place of their own, these projects reliably failed to stimulate sustained economic development and security within host communities. Worse yet, the projects often exacerbated community *disempowerment* by fostering members' disengagement with community problem solving, even in ostensibly participatory development projects (Cooke and Kothari, 2001).

Lucena et al. recognize that "the idea of making 'community' central to engineering for development is more complex than it sounds" (2010, 5), and then go on to discuss methods and approaches that better integrate community members into engineers' sustainable development projects. In taking this approach, work in engineering for sustainable community development directs much-needed attention to the first of the 3 Ps: people, which is arguably the most important yet least well understood element in engineering-for-development projects. While the tools for assessing the economic benefit of technologies are often ready-at-hand, and the tools for assessing environmental impacts are increasingly available, engineering for sustainable community development work addresses conspicuous deficiency in the tools for assessing and ensuring *sustained social* benefits derive from engineers' development interventions.

This shift in attention to community in engineering-for-development work offers promise, even as many significant problems in practice remain (notably, "voluntourism" and woefully underprepared volunteer "experts"). One particular area of promise in the shift in attention is that it highlights the social power relations at play within the contexts of development work, both locally and internationally. This challenges engineers to confront those power relations in productive ways, both to the benefit of community members and in relation to engineering's professional ecology. Such attention also results in better learning opportunities for both volunteer engineers and community members, which is a positive outcome in itself, regardless of the nature or even the "success" of the implemented technology.

For the most part, engineers advocating or working on sustainable community development projects do not take an explicitly anti-militaristic stance; in fact, they frequently say little regarding militarism at all. And yet the contexts of application of engineering expertise, the end goals of the work, and the problem-solving approaches applied all offer compelling alternatives to dominant

---

[25] Here, we point to projects carried out by numerous distinct non-profit organizations and educational institutions, whether or not affiliated with any particular Engineers Without Borders organization. (Many, independent Engineers Without Borders organizations exist internationally.)

models of engineering—both those taught in most educational settings and those typically practiced within the military-industrial-academic complex.

In particular, this work concretizes its commitment to *community empowerment* and seeks to cultivate the practices and insights engineers need to serve communities on their own terms, a compelling alternative to developing technologies of social control. Like the engineering for peace and justice work, these initiatives change the conceptual terrain of engineering—expanding imagination for what engineering is, what engineers do, and what expertise and skills they require to be effective in advancing a broader vision of "the common good."

Engineering for peace and justice and engineering for sustainable community development each offers models and strategies for addressing the inordinate influence of militarism on engineering even where militarism is not explicitly addressed. More work certainly needs to be done in these domains to identify specific strategies for responding to the evolving nature of military-technology development and the present configuration of the military-industrial-academic complex. Yet the theoretical and practice-oriented tools needed to carry out this analysis are already in place within these "movements." Common across them are the rejection of *the myth of engineering's neutrality* and the embrace of explicit *politics of empowerment*, particularly for marginalized social groups—including sometimes those at the receiving end of technologies of control and surveillance.

We see the best of these approaches upsetting the traditional modes of patronage and subservience that have characterized so much of engineering over its history. Without romanticizing "the local," they find an alternative to the large corporation, military or otherwise, and the nation-state, whose bidding engineers have traditionally done. Moreover, they tend to embrace a conception of community that is at once both much smaller as well as much broader than the nation-state or the transnational firm's stockholders. They focus on the complex processes for addressing specific communities' specific needs, and simultaneously locate themselves among many such communities, each of which is beset by economic marginalization and political exclusion, in many cases derived from the long legacies of colonialism.

CHAPTER 7

Conclusion: Facing the Entwinement of Engineering and War

This book has explored three different avenues for engineers and engineering students to think through their careers with a rich sense of integrity and of history, as well as an open understanding of what the future can bring. Investigating engineering and warfare brings into focus complex and challenging issues, some of which may be well beyond the scope of traditional approaches to engineering education and practice. But we hope that these tools from the humanities and the social sciences—history, philosophy, critical theory, and technology studies—can be wedded to engineering pedagogies and practices in a way that helps engineers throughout their professional lives.

The connections between engineering and warfare have changed over time, in relation to the goals of states and private firms as well as in relation to more progressive social movements. States' primary political goals, traditionally, were to acquire new territories and access to new markets. This was based around the concept of imperial expansion in which dominant (usually European) powers would conquer and then control colonial possessions in the global South, installing friendly regimes and extracting resources and wealth to put into the metropolitan coffers. The means of that acquisition was through weaponry that could kill, maim, or destroy opponents. Those weapons employed physical, kinetic force, and, primarily, the application of Newtonian physics to human bodies. This period saw the technology-driven intensification of the application of force via the development of the industrial and later the chemical revolutions. The period ended, for the most part, with decolonization movements in the middle years of the 20th century, though its principal weapons have hardly disappeared and remain in much use.

As we have seen, the Second World War was a turning point in world history. New military technologies—from radio to radar and nuclear weaponry—took hold and helped determined the war's outcome. Because of the astounding technological sophistication of these devices, university-based researchers and private funders played key roles in weapons development. This relationship was enormously profitable for many businesses and for many academics; it was also a tremendously effective way to develop weapons of unprecedented power. Moreover, it helped give rise to what U.S. President Dwight Eisenhower—a proponent of this collaboration for most of his adult life—would famously call the military-industrial complex. The Second World War so weakened European powers—the British, Germans, and French especially—that the existing imperial world economic system was undone. Anti-imperial nationalists across Africa, Asia, and Latin America claimed their own lands and resources for their own peoples. The European powers could no longer forcibly restrain anti-imperial movements across the colonial world, and movements of national liberation swept the southern hemisphere.

At the same time, the United States and the Soviet Union engaged as the main combatants in the Cold War, an effort to influence the world that would follow from the one 19th-century colonialism had wrought. They waged arms races and developed their own, twinned military-industrial complexes. They still fought for territories using kinetic weapons systems, but also using rival political and economic ideologies. The nuclear doctrine of Mutually Assured Destruction guaranteed that the battles would not take place on the home soil of either country. Instead, the Cold War was fought through a series of proxy wars, hot and cold, across the formerly colonized Third World. Propaganda battles for hearts and minds were among the new methods of military struggle, and universities became sites in this battle, too.

As the terrain of geopolitics, economics, and technologies shifted in the late-20th and early-21st centuries, so too did the terrain of war. The traditional lines between internal and external enemies blurred, with increased surveillance and "soft-kill" technologies called upon as a response to the threat of domestic unrest and political violence. With the proliferation of drone and robotic warfare, naked violence and bloodshed have remained. While it is too soon to know with certainty, we fear that the United States' engagement in an endless War on Terror may become paradigmatic for nation-states across the planet.

If that becomes the case, the U.S. Defense Advanced Research Projects Agency's spiral of innovation will radically intensify, taking place in labs and universities around the world. The allure of cutting-edge engineering research and generous funding, and the trickle down effect it has on universities—as well as the blowback of unintended uses of these fearsome and subtle technologies, from destruction and pain to surveillance and informational control—will continue producing new dreams, terrors, nightmares, and opportunities.

But it is not only the particular trajectories taken by military technologies that should be of concern to engineers as they imagine their profession's roles and navigate their careers. It is the entirety of the military-industrial-academic complex that confronts contemporary citizenry, and that confronts engineers with particular stridency. The close alignments among military agencies, the defense industry, and government-funded academic research in science and engineering come together as a system such that any one individual in the system cannot easily or straightforwardly "opt out," as was promoted by some of the 1970s reformers. Identifying career paths with integrity, the thrust of our analysis suggests, requires engineers to face the extensive interconnections between, on one hand, engineering as a field—both academic and in industry—and, on the other hand, military interests and agendas.

To say that engineers are inevitably implicated in larger systems that work to advance militarism is not to suggest they cannot identify career paths with integrity. First, as we stated above, we do not assume careers within the military are necessarily without integrity. In some instances and for some individuals, careers developing military technologies may meet the requirements of individual integrity described in Chapter 3. Second, and more important to the thrust of our analysis, this book seeks to highlight that acting with integrity requires confronting the conceptual and institutional frameworks—the social, political, and economic structures that constrain one's space of movement—and how those frameworks are likely to shape and direct one's work in ways that transcend, perhaps even contradict, one's own intentions and commitments.

In this way, acknowledgement of—and some degree of critical engagement with—the military-industrial-academic complex is a basic requirement for all engineers (as well as for many other experts and occupational groups) seeking to go about their work with integrity. Similarly, given the pervasive influence of the military-industrial-academic complex on engineering, historically and in the present, and given the numerous ways evolving military strategies have resulted in entirely new technology research initiatives, engineering reformers would be amiss to ignore militarism as a central shaper of the field.

As we have shown throughout our analysis, this influence is both broad and deep. Broad in the sense of shaping much of what constitutes engineering, from how it is practiced within hierarchical organizations to its thoroughgoing separation of technical and social dimensions to its masculinist assumptions and values. Deep in the sense that so very much of engineering activity and expertise is directed at military problems, whether considering warfare itself or more generally the problems of surveillance and control.

Students and educators alike often contrast students' sometimes-insulated lives in university with the host of obligations and responsibilities that follow once they graduate, when they enter what is often referred to as "the real world." There, what is important is a well paying job, financial security, achieving success in a world of hard knocks. But this real world is a simplification, too, ignoring as it does the responsibilities individuals, professionals, experts have toward themselves and toward the public welfare. As university graduates enter working life, they should maintain no illusion that they can individually surmount enduring alignments between engineering and militarism. Yet neither should they seek to abrogate responsibility for identifying career paths with integrity by claiming impotency in the face of systemic challenges. The working life of engineers must take place in the social, political, and personal milieus that constitute the world as it actually is—what we might call the "*real* real world."

As we return to one of the definitions of engineering offered early in the book, we hope that new insights and definitions might emerge.

> Engineering is the science, skill, and profession of acquiring and applying scientific, economic, social, and practical knowledge, in order to design and also build structures, machines, devices, systems, materials and processes.[26]

---

[26] Wikipedia. The Free Encyclopedia. First sentence of entry on "Engineering." Accessed November 17, 2012 at http://en.wikipedia.org/wiki/Engineering.

Engineers work at the interface of the human and the technical, revealing the porousness of the line between them. As individuals, as a collective, as a social movement, engineers will play a crucial role in making the world that is to come.

Regardless of one's definition of engineering, the question of what engineers do will remains of vital importance. Engineering is an eminently social practice with profound ramifications on social life, but always within historical and geopolitical contexts that influence the possibilities of human action. Engineers are called upon to make difficult decisions not just about technical solutions, but about the nature of political life and social conflict, about the ethics of warfare and its increasingly tangled lines of accountability and responsibility. As our understandings of biological, sociological, and technological systems transform and intertwine, we believe that, perhaps as never before, engineers should be equipped with an understanding of the historical, political, and ethical consequences of the work they do and the lives they lead.

Additional Resources

The website of the Engineering, Social Justice, and Peace network: http://esjp.org

The website of the *International Journal of Engineering, Social Justice, and Peace*: http://library.queensu.ca/ojs/index.php/IJESJP

Morgan and Claypool Publishers series *Synthesis Lectures on Engineers, Technology and Society*, edited by Caroline Baillie: http://www.morganclaypool.com/toc/ets/6/3

Nieusma and Blue's 2012 "Engineering and War" article, *IJESJP* 1(1): 50–62

C. Richard Bowen's 2009 *Engineering Ethics: Outline of an Aspirational Approach*: http://www.springer.com/engineering/book/978-1-84882-223-8

Vesilind's *Peace Engineeing: When Personal Values and Engineering Careers Converge*, 2nd Edition (2013); P. Aarne Vesilind and W. Richard Bowen, editors; Lakeshore Press: http://lakeshorepressbooks.com/index.php/store/category/1

References

Alcoff, L. M. (2002). "Does the Public Intellectual Have Intellectual Integrity?" *Metaphilosophy* 33: 521–534. DOI: 10.1111/1467-9973.00246.

Althusser, L. (1971). *Lenin and Philosophy and Other Essays*, Ben Brewster (trans.), New York: Monthly Review Press.

Aquinas, T. (1265-1274). *Summa Theologica*.

Arike, A. (2010 March). "The Soft Kill Solution: New Frontiers in Pain Compliance." *Harper's*, 38-47.

Ashford, E. (2000). "Utilitarianism, Integrity and Partiality." *Journal of Philosophy* 97: 421–439. DOI: 10.2307/2678423.

Babbitt, S. E. (1997). "Personal Integrity, Politics and Moral Imagination." Brennan, S; Isaacs, T. and Milde, M. (eds.). *A Question of Values: New Canadian Perspectives on Ethics and Political Philosophy*. Amsterdam and Atlanta: Rodopi, 107–31.

Bailey, S. (1972). *Prohibitions and Restraints in War*. Oxford: Oxford University Press. DOI: 10.1017/S0035336100110597.

Baillie, C. (2006). *Engineers within a Local and Global Society*. Morgan and Claypool Publishers. DOI: 10.2200/S00339ED1V01Y201104ETS016.

Baillie, C., and Levine, M. (2013). "Engineering Ethics from a Justice Perspective: A Critical Repositioning of What It Means To Be an Engineer." *International Journal of Engineering, Social Justice, and Peace*, 2(1), 10-20.

Baudrillard, J. (1995). *The Gulf War Did Not Take Place*. Bloomington: Indiana University Press.

Bergen, P., & Pandey, S. (2006). "The Madrassa Scapegoat." *Washington Quarterly*, 29(2), 115-125. DOI: 10.1162/wash.2006.29.2.117.

Best, G. (1980). *Humanity in Warfare*. New York: Columbia University Press, 1980.

Blustein, J. (1991). *Care and Commitment: Taking the Personal Point of View*. New York: Oxford University Press. DOI: 10.1093/acprof:oso/9780195067996.001.0001.

Bourne, R. (1918). "The State," in *War and the Intellectuals: Collected Essays, 1915-1919*, Carl Resek (Ed). Indianapolis: Hackett, 1964.

Bowen, W. R. (2009). *Engineering Ethics: Outline of an Aspirational Approach*. London: Springer. DOI: 10.1007/978-1-84882-224-5.

Brown, W. (2003). "Neo-liberalism and the End of Liberal Democracy." *Theory and Event* 7: 1. http://muse.uq.edu.au/journals/theory_and_event/v007/7.1brown.html. DOI: 10.1353/tae.2003.0020.

Brown, W. (2006a). 'American Nightmare: Neoliberalism, Neoconservatism and De-Democratization', *Political Theory*, 34: 6, 690–714. DOI: 10.1177/0090591706293016.

Brown, W. (2006b). *Regulating Aversion: Tolerance in the Age of Identity and Empire*. Princeton: Princeton University Press. DOI: 10.1017/S1537592707070855.

Brundtland, G. H. (1987). *Report of the World Commission on Environment and Development: "Our Common Future."* United Nations.

Bucciarelli, L. L. (1994). *Designing Engineers*. Cambridge, Massachusetts: MIT Press. DOI: 10.1080/03043799508928289.

Butler, J. (1999). *Gender Trouble: Feminism and the Subversion of Identity*. New York: Routledge.

Calhoun, C. (1995). "Standing for Something." *Journal of Philosophy*, XCII, 235–260. DOI: 10.2307/2940917.

Carnegie Foundation. (1984). "National Survey of Higher Education." The Association of Religion Data Archives' webpage, www.thearda.com/Archive/Files/Descriptions/NSHEF84.asp.

Catalano, G. (2006). *Engineering Ethics: Peace, Justice, and the Earth*. Morgan and Claypool. (www.morganclaypool.com). DOI: 10.2200/S00039ED1V01Y200606ETS001.

Chafe, W.H. (2010). *The Unfinished Journey: America Since World War II*, 7th ed. New York: Oxford University Press.

Chambers II, J. W. ed. (1999). *The Oxford Companion to American Military History*. New York: Oxford University Press.

Christensen, S. H., Delahousse, B., and Meganck, M., Editors. (2009). *Engineering in Context*. Aarhus, Denmark: Academica.

Clausewitz, C. von. (1995). *On War*, trans. by A. Rapoport. Harmondsworth, UK: Penguin.

Cockburn, C., and Ormrod, S.. (1993). *Gender and Technology in the Making*. Thousand Oaks, California: Sage.

Connell, R. W. (1995). *Masculinities*. Berkeley: University of California Press.

Cooke, B. and Kothari, U., Editors. (2001). *Participation: The New Tyranny?* London: Zed Books.

Cooling, B.F., ed. (1981). *War, Business, and World Military-Industrial Complexes.* Port Washington, NY: National University Publications, 1981.

Cox, D., La Caze, M. and Levine, M.(1999). "Should We Strive for Integrity?" *Journal of Value Inquiry* 33/4, 519–530. DOI: 10.1023/A:1004614232579.

Cox, D., M. La Caze, M. Levine (2003). *Integrity and the Fragile Self.* London: Ashgate.

Cox, D. (2005). "Integrity, Commitment, and Indirect Consequentialism." *Journal of Value Inquiry* 39, 61-73. DOI: 10.1007/s10790-006-1571-7.

CSRE—Committee for Social Responsibility in Engineering. 1971. "Statement of Purpose." *SPARK* vol. 1, no. 1, p. 3.

CSRE—Committee for Social Responsibility in Engineering. 1973. "The Forging of an Engineer's Conscience." *SPARK* vol. 3, no. 2, pp. 2-5.

DARPA – Defense Advanced Research Projects Agency. (2012a). "Success Reports" webpage. Accessed 28 March 2012 at: http://www.darpa.mil/Opportunities/SBIR_STTR/SBIR_ STTR_Success_Reports.aspx.

DARPA – Defense Advanced Research Projects Agency. (2012b). "News and Events/Information," DARPA webpage, http://www.darpa.mil/our_work/, accessed March 28 2012.

DARPA – Defense Advanced Research Projects Agency. (2012c). "What Keeps DARPA Leadership up at Night," DARPA webpage, http://www.darpa.mil/NewsEvents/Releases/2012/02/29a.aspx accessed online March 28 2012.

DARPA – Defense Advanced Research Projects Agency. (2012d). "Department of Defense Fiscal Year (FY) 2013 President's Budget Submission," Defense Advanced Research Projects Agency, *Justification Book Volume 1: Research, Development, Test & Evaluation, Defense-Wide* (unclassified) Feb 2012.

DARPA – Defense Advanced Research Projects Agency. (2002). Fiscal Year (FY) 2003 Budget Estimates, February 2002, *Research, Development, Test, and Evaluation, Defense-Wide Volume 1 - Defense Advanced Research Projects Agency* (unclassified)

Davion, V. (1991). "Integrity and Radical Change." *Feminist Ethics.* Ed. Claudia Card, Lawrence, Kansas: University of Kansas Press, 180–192.

Dickson, D. (1974). *The Politics of Alternative Technology.* New York: Universe Books.

Dunn, P. D. (1978). *Appropriate Technology: Technology with a Human Face.* London: Macmillan.

Dunn, T.J. (1995). *The Militarization of the U.S.-Mexico Border: Low-Intensity Conflict Doctrine Comes Home.* Austin: CMAS Books.

Edwards, P. N. (1996). *The Closed World: Computers and the politics of discourse in Cold War America.* Cambridge: MIT Press.

ESJP – Engineering, Social Justice, and Peace. (2010). "Our Commitments." Accessed 19 June 2013 at: http://esjp.org/about-esjp/our-commitments.

Faure-Brac, R. (2012). *Transition to Peace: A Defense Engineer's Search for an Alternative to War.* iUniverse.

Ferguson, E. S. (1992). *Engineering and the Mind's Eye.* MIT Press.

Foucault, M. (1979). *Discipline and Punish: The Birth of the Prison.* Trans Alan Sheridan. New York: Vintage.

Foucault, M. (1990). *The History of Sexuality, Vol. I. An Introduction.* Trans Robert Hurley. New York: Vintage.

Frankfurt, H. (1987). 'Identification and Wholeheartedness.' Ferdinand Schoeman, ed. *Responsibility, Character, and the Emotions: New Essays in Moral Psychology*, New York: Cambridge University Press. DOI: 10.2307/2381243.

Freud, S. (1919). "The Uncanny." *The Standard Edition of the Complete Psychological Works of Sigmund Freud*, 17, 219-52.

Fukuyama, F. (1992). *The End of History and the Last Man.* New York: Free Press.

Fulbright, J. W. (1970). "The War and Its Effects: The military-industrial-academic complex." *Super State: Readings in the military-industrial complex*, Schiller, H. I., & Phillips, J. D. (Eds). Urbana: University of Illinois Press, 171-178.

Gambetta, D., and Hertog, S. (2009). "Why Are There So Many Engineers among Islamic Radicals?" *European Journal of Sociology*, 50(2), 201-230. DOI: 10.1017/S0003975609990129.

Gilbert, M. (1997). "Group Wrongs and Guilt Feelings." *The Journal of Ethics* 1, 65-84 DOI: 10.1023/A:1009712003678.

Gilligan, C. (1982). *In a Different Voice.* Cambridge, Mass: Harvard University Press.

Giroux, H.A. (2007). *The University in Chains: Confronting the Military-Industrial-Academic Complex.* Boulder, CO: Paradigm Publishers.

Graham, J. L.(2001). "Does Integrity Require Moral Goodness?" *Ratio* 14: 234–251. DOI: 10.1111/1467-9329.00160.

Gramsci, A. (1971). *Selections from the Prison Notebooks*, ed and trans by Quintin Hoare and Geoffrey Nowell Smith. New York: International Publishers.

Grant, R. W. (1997). *Hypocrisy and Integrity.* Chicago and London: University of Chicago Press.

Guy, S. (2013a). "Women in Military Labs Contribute on a Multitude of Fronts," *SWE: Magazine of the Society of Women Engineers*, Vol. 59, No. 1, 36-40.

Guy, S. (2013b). "Women in Military Labs Share Personal Stories and Insights," *SWE: Magazine of the Society of Women Engineers*, Vol. 59, No. 1, 40a-40d.

Hacker, S. (1989). *Pleasure, Power, and Technology: Some tales of gender, engineering, and the cooperative workplace*. Boston: Unwin Hyman.

Halfon, M. (1989). *Integrity: A Philosophical Inquiry*, Philadelphia: Temple University Press.

Hambling, D. (2008 July 3). "Microwave Ray Gun Controls Crowds with Noise," *NewScientist*. Accessed 13 April 2012 at: http://www.newscientist.com/article/dn14250-microwave-ray-gun-controls-crowds-with-noise.html.

Harris, G. W. (1989). "Integrity and Agent Centered Restrictions." *Nous* 23: 437–456. DOI: 10.2307/2215877.

Harvey, D. (1990). *The Condition of Postmodernity: An Inquiry into the Origins od Cultural Change*. Cambridge, MA: Blackwell.

Harvey, D. (2007). *A Brief History of Neoliberalism*. New York: Oxford University Press.

Hebert, M. R. (2002). "Integrity, Identity and Fanaticism." *Contemporary Philosophy* 24: 25–29.

Held, D. (1996). "The Development of the Modern State," in *Modernity: An Introduction to Modern Societies*, Stuart Hall, David Held, Don Hubert, and Kenneth Thompson (eds). Cambridge, MA: Blackwell, 55-89.

Herman, B. (1983). "Integrity and Impartiality." *Monist* 66: 233–250. DOI: 10.5840/monist198366216.

Higham, R. (1981). "Complex Skills and Skeletons in the Military-Industrial Relationship in Great Britain." *War, Business, and World Military-Industrial Complexes*, Benjamin Franklin Cooling, (Ed). Port Washington, NY: National University Publications, 8-32.

Holley, D. M. (2002). "Self-Interest and Integrity." *International Philosophical Quarterly*, 42: 5–22. DOI: 10.5840/ipq200242170.

Homze, E. M. (1981). "The German MIC." *War, Business, and World Military-Industrial Complexes*, Cooling, B. F. (Ed). Port Washington, NY: National University Publications, 51-83.

Ikegami-Andersson, M. (1992). *The Military-Industrial Complex: The Cases of Sweden and Japan*. Aldershot, UK: Dartmouth Publishing.

Jeffreys, D. (2008). *Hell's Cartel: IG Farben and the Making of Hitler's War Machine*. London: Bloomsbury.

Johnson, J. T. (1984). *Can Modern War Be Just?* New Haven, CT: Yale University Press.

Johnson, J. T. (1981). *Ideology, Reason and Limitation of War: Religious and Secular Concepts*, 1200-1740. Princeton, NJ: Princeton University Press.

Johnson, J. T. (1999). *Morality and Contemporary Warfare*. New Haven, CT: Yale University Press.

Johnson, J. T. (1981). *The Just War Tradition and the Restraint of War*. Princeton, NJ: Princeton University Press.

Johnson, J. T. (1973). "Towards Reconstructing the *Jus ad Bellum*", *Monist*, 461-88. DOI: 10.5840/monist197357422.

Khatchadourian, R. (2012 December 17). "Operation Delirium." *The New Yorker*, 46-64.

Kunda, G. (1992). *Engineering Culture: Control and Commitment in a High-tech Corporation*. Temple University Press.

Law, J. (1987). "Technology and Heterogeneous Engineering: The Case of Portuguese Expansion." *The Social Construction of Technological Systems: New Directions in the Sociology and History of Technology*, 111-134.

Layton, E. T. (1986). *The Revolt of the Engineers: Social Responsibility and the American Engineering Profession*. Johns Hopkins University Press.

Layton, E. T. (1983). "Engineering Needs a Loyal Opposition: An Essay Review." *Business & Professional Ethics Journal*, 2(3), 51-59. DOI: 10.5840/bpej19832324.

Leslie, S. W. (1993). *The Cold War and American Science: The Military-Industrial-Academic Complex at MIT and Stanford*. New York: Columbia University Press.

Lipsitz, G. (1994). *Rainbow at Midnight: Labor and Culture in the 1940s*. Urbana: University of Illinois Press.

Lucena, J. C. (2005). *Defending the Nation: U.S. Policymaking to Create Scientists and Engineers from Sputnik to the "War Against Terrorism"* Lanham: University Press of America.

Lucena, J., Schneider, J., and Leydens, J. A. (2010). *Engineering and Sustainable Community Development*. Morgan and Claypool.

Martin, W. G. (2005 Spring). "Manufacturing the Homeland Security Campus and Cadre." *ACAS [Association of Concerned Africa Scholars] Bulletin* 70: 27-32. Accessed 3 July 2013 at: http://concernedafricascholars.org/docs/acasbulletin70.pdf.

Masco, J. (2006). *The Nuclear Borderlands: The Manhattan Project in Post-Cold War New Mexico*. Princeton: Princeton University Press.

McCuen, H. and K. Gilroy (2010). *Ethics and Professionalism in Engineering*. Peterborough, Ontario: Broadview.

McFall, L. (1987). 'Integrity.' Ethics 98, 5–20. Reprinted in John Deigh (ed.), *Ethics and Personality*, Chicago: University of Chicago Press, 1992, 79–94.

Meiksins, P. (1996). "Engineers in the United States: A House Divided." *Engineering Labour: Technical Workers in Comparative Perspective*, 61-97.

Meiksins, P. F., and Smith, C., Editors. (1996). *Engineering Labour: Technical Workers in Comparative Perspective*. Verso Books.

Monk, Ray. (2012). *Inside the Centre: The Life of J. Robert Oppenheimer*. London: Jonathan Cape.

Moore, K. (2008). *Disrupting Science: Social Movements, American Scientists, and the Politics of the Military, 1945-1975*. Princeton University Press.

Narveson, J. (2002). "Collective Responsibility." *The Journal of Ethics* 5, 105-120.

Nieusma, D. (2013). "Engineering, Social Justice, and Peace: Strategies for Educational and Professional Reform." In *Engineering Education for Social Justice* (pp. 19-40). Springer Netherlands. DOI: 10.1007/978-94-007-6350-0_2.

Nieusma, D., and Blue, E. (2012). Engineering and War. *International Journal of Engineering, Social Justice, and Peace*, 1(1), 50-62.

Nieusma, D., and Riley, D. (2010). Designs on Development: Engineering, Globalization, and Social Justice. *Engineering Studies*, 2(1), 29-59. DOI: 10.1080/19378621003604748.

Noble, D. F. (1979). *America by Design: Science, Technology, and the Rise of Corporate Capitalism*. Oxford University Press.

Orend, B. (2000). "War." http://plato.stanford.edu/entries/war/. First published Feb 4, 2000; revised Jul 28, 2005.

Ortner, Sherry B. (1996). *Making Gender: The Politics and Erotics of Culture*. Boston: Beacon Press.

Papadopoulos, C., and Hable, A. (2008). "Including Questions of Military and Defense Technology in Engineering Ethics Education." *Proceedings of the American Society for Engineering Education Annual Conference and Exposition*.

Parenti, C. (1999). *Lockdown America: Police and Prisons in the Age of Crisis*. London: Verso.

Pawley, A. L. (2012). "What Counts as 'Engineering': Toward a redefinition." *Engineering and Social Justice: In the university and beyond*, 59-85.

Petrillo, A.M. (2011 May 5). "Miltech—Flying Snakes Get a Close Look by DARPA." Military Officer's Association of America. Accessed 28 March 2012 at: http://moaablogs.org/message/2011/05/miltech-%E2%80%94-flying-snakes-get-a-close-look-by-darpa/.

Phillips, R. (1984). *War and Justice*. Norman Oklahoma: University of Oklahoma Press. DOI: 10.1080/07418828400088161.

Pilisuk, M., and Hayden, T. (1965). "Is There a Military Industrial Complex Which Prevents Peace? Consensus and Countervailing Power in Pluralistic Systems." *Journal of Social Issues*, 21(3), 67-117. DOI: 10.1111/j.1540-4560.1965.tb00506.x.

Popper, B. (2009, December). "Build-a-Bomber: Why Do So Many Terrorists Have Engineering Degrees?" *Slate Magazine*.

Pursell, Jr, Carroll W. (1972). *The Military-Industrial Complex*. New York: Harper and Row.

Putman, D. (1996). "Integrity and Moral Development." *The Journal of Value Inquiry*, 30: 237–246.

Remarque, E. M. (1929). *All Quiet on the Western Front*. DOI: 10.1007/BF00162894.

Reynolds, T. S. (Ed.). (1991). *The Engineer in America: A Historical Anthology from Technology and Culture*. University of Chicago Press.

Riley, D. (2008). *Engineering and Social Justice*. Morgan and Claypool.

Roland, A. (2001). *The Military-Industrial Complex*. Washington, DC: American Historical Association.

Rostow, W. W. (1960). *The Stages of Economic Growth: A Non-Communist Manifesto*. Cambridge: Cambridge University Press.

Rothman, S., Lichter, S. R., & Nevitte, N. (2005, March). "Politics and Professional Advancement among College Faculty." *The Forum* 3(1): 1-16.

Sandel, M. (2009). *Justice: What's the Right Thing to Do?* New York: Farrar, Strauss, Giroux.

Schiller, H.I., ed. (1970). *Super-State: Readings in the Military-Industrial Complex*. Urbana: University of Illinois Press.

Schlosser, E. (2013). *Command and Control: Nuclear Weapons, the Damascus Incident, and the Illusion of Safety*. New York: Penguin.

Schumacher, E. F. (1973). *Small Is Beautiful: Economics As If People Mattered*. Harper Perennial.

Schwartz, C. (2011). A Partial Archive of Science for the People: Contents – Timeline. Accessed 30 May 2013 at: http://socrates.berkeley.edu/~schwrtz/SftP/Contents.html. DOI: 10.1039/c0py00246a.

Schweber, S. S. (1988). "The Mutual Embrace of Science and the Military: ONR and the growth of physics in the United States after World War II." *Science, Technology and the Military*, Volume 1, 1-45.

Seagrave, S. (November 9, 1964). "Play about Him Draws Protests of Oppenheimer." *The Washington Post*: B8.

SESPA – Scientists and Engineers for Social and Political Action, Berkeley Chapter (1972). "Science against the People." Self-published report.

Seymour, E., and Hewitt, N. M. (1997). *Talking about Leaving: Why undergraduates leave the sciences*. Boulder, CO: Westview Press.

Simpson, C. (1998). "Universities, Empire, and the Production of Knowledge: An Introduction." *Universities and Empire: Money and Politics in the Social Sciences During the Cold War*, Simpson, C. (Ed). New York: New Press.

Smillie, I. (2000). *Mastering the Machine Revisited: Poverty, Aid and Technology*. London: ITDG Publishing.

Spaulding, C. B., & Turner, H. A. (1968). "Political Orientation and Field of Specialization among College Professors. *Sociology of Education*, 247-262. DOI: 10.2307/2111874.

SPPA – Scientists for Social and Political Action. (1969a). Call for Participation. Accessed 30 May 2013 at: http://socrates.berkeley.edu/~schwrtz/SftP/SSPA1.html.

SPPA – Scientists for Social and Political Action. (1969b). Newsletter 1, 23 February 1969. Accessed 30 May 2013 at: http://socrates.berkeley.edu/~schwrtz/SftP/SSPA2.html.

Stearns, P. N. (1979). *Be a Man! Males in Modern Society*. New York: Holmes and Meier.

Streett, W. B. (1993). "The Military Influence on American Engineering Education." *Cornell Engineering Quarterly*. 27(2), 3-10.

Sutherland, S. (1996). "Integrity and Self-Identity." *Philosophy*, Supplementary Vol. 35: 19–27.

Swett, C. (1993). "Strategic Assessment: Non-Lethal Weapons," Office of the Assistant Secretary of Defense for Special Operations and Low-Intensity Conflict.

Taylor, P. W. (1975). *Principles of Ethics: An Introduction*. Belmont, California: Wadsworth Publishing.

Taylor, G. (1981). "Integrity." *Proceedings of the Aristotelian Society*, Supplementary Vol. 55: 143–159.

Taylor, G. (1985). "Integrity." *Pride, Shame and Guilt: Emotions of Self-Assessment*. Oxford: Oxford University Press, pp. 108–141.

Teichman, J. (1986). *Pacifism and the Just War*. Oxford: Basil Blackwell.

Tonso, K. L. (2007). *On the Outskirts of Engineering: Learning identity, gender, and power via engineering practice*. Sense Publishers.

Trianosky, G. W. (1986). "Moral Integrity and Moral Psychology: A Refutation of Two Accounts of the Conflict Between Utilitarianism and Integrity." *Journal of Value Inquiry* 20, 279–288. DOI: 10.1007/BF00146117.

Tucker, R. (1960). *The Just War: A Study in Contemporary American Doctrine*. Baltimore, MD: Johns Hopkins University Press.

Unger, S. H. (2013). "Review of *Engineers for Change: Competing Visions of Technology in 1960s America*, Matthew Wisnioski." *Social Epistemology Review and Reply Collective* 2 (6): 1-4.

Van de Poel, I. and L. Royakkers (2011). *Ethics, Technology and Engineering: An Introduction*. West Sussex: Wiley-Blackwell.

Van Hooft, S. (2001). "Judgment, Decision, and Integrity." *Philosophical Explorations* 4: 135–149. DOI: 10.1080/10002001058538712.

Vesilind, P. A. (2010). *Engineering Peace and Justice: The Responsibility of Engineers to Society*. Springer.

Vesilind, P. A., ed. (2005). *Peace Engineering: When Personal Values and Engineering Careers Converge*. Lakeshore Press.

Vincenti, W. G. (1990). *What Engineers Know and How They Know It: Analytical studies from aeronautical history*. John Hopkins University Press.

Vinck, D. (2003). *Everyday Engineering: An Ethnography of Design and Innovation*. MIT Press.

Virilio, P. (2007 [1977]). *Speed and Politics: An Essay on Dromology*. Trans M. Polizzotti. New York, Semiotext(e).

Waheed, S. (2012 January 3). "The Malalas You Don't See." The Nation. Accessed 3 January 2013 at: http://www.thenation.com/article/170920/malalas-you-dont-see#axzz2XM0KtVwX.

Walzer, M. (2004). *Arguing About War*. New Haven: Yale University Press.

Walzer, M. (2000). *Just and Unjust Wars: A Moral Argument with Historical Illustrations*. New York: Basic Books, 3rd ed.

Walzer, M. (1970). *Obligations: Citizenship, War and Disobedience*. Harvard: Harvard University Press.

Walzer, M. (1990). "Nation and Universe" in G.B. Peterson, ed. *The Tanner Lecture on Human Values*. Salt Lake City, Utah: Utah University Press, pp. 507-56.

Wasserstrom, R. (1971-72). "The Relevance of Nuremberg," *Philosophy and Public Affairs* 22-46.

Wasserstrom, R. ed. (1970). *War and Morality*. Belmont, CA: Wadsworth.

Wells, D.A. (1996). *An Encyclopedia of war and ethics*. Westport, CT: Greenwood.

Williams, B. (1973). "Integrity." In J.J.C. Smart and Bernard Williams, *Utilitarianism: For and Against*. New York: Cambridge University Press, 108-117.

Williams, B. (1981a). 'Utilitarianism and Moral Self-Indulgence.' In *Moral Luck: Philosophical Papers 1973-1980*. Cambridge: Cambridge University Press, 40-53. DOI: 10.1017/CBO9781139165860.004.

Williams, B. (1981b). 'Persons, Character and Morality.' In *Moral Luck: Philosophical Papers 1973-1980*. Cambridge: Cambridge University Press, 1-19. DOI: 10.1017/CBO9781139165860.002.

Williams, B. (1981c). 'Moral Luck.' In *Moral Luck: Philosophical Papers 1973-1980*. Cambridge: Cambridge University Press, 20-39. DOI: 10.1017/CBO9781139165860.003.

Williams, B. (1985). *Ethics and the Limits of Philosophy*. Cambridge: Harvard University Press.

Willoughby, K. W. (1990). *Technology Choice: A Critique of the Appropriate Technology Movement*. Westview Press. DOI: 10.1002/jid.3380050312.

Wisnioski, M. H. (2009). "How Engineers Contextualize Themselves." *Engineering in Context*, 403-415.

Wisnioski, M. H. (2012). *Engineers for Change: Competing Visions of Technology in 1960s* America. MIT Press.

Yoder, J. (1984). *When War is Unjust: Being Honest in Just-War Thinking*. Minneapolis, Augsburg Press.

Zagzebski, L. (1996). *Virtues of the Mind: An inquiry into the nature of virtue and the ethical foundations of knowledge*. Cambridge: Cambridge University Press. DOI: 10.1017/CBO9781139174763.

Author Biographies

Ethan Blue is an associate professor of history at the University of Western Australia. He is the author of *Doing Time in the Depression: Everyday Life in Texas and California Prisons* (New York University Press, 2012), with additional publications in *Pacific Historical Review; Law, Culture, and the Humanities; Journal of Social History;* and, with Dean Nieusma, the *International Journal of Engineering, Social Justice, and Peace.* He is currently researching the historical technologies of American deportation and the comparative history of settler colonialisms.

Michael Levine is a professor of philosophy at the University of Western Australia. He is the author of the following books: *Prospects for an Ethics of Architecture*, with Bill Taylor (2011); *Doing Philosophy, Watching Movies*, with Damian Cox (2011); *Politics Most Unusual: Violence, Sovereignty and Democracy in the "War on Terror,"* with Damian Cox and Saul Newman (2009); *Integrity and the Fragile Self*, with Damian Cox and Marguerite La Caze (2003); and *Pantheism: A Non-theistic Concept of Deity* (1994). Levine has also edited *Racism in Mind*, with Tamas Pataki (2004), and *The Analytic Freud: Philosophy and Psychoanalysis* (2000).

Dean Nieusma is an associate professor of science and technology studies at Rensselaer Polytechnic Institute, where he is Director of Rensselaer's acclaimed Programs in Design and Innovation. He is founding editor of the *International Journal of Engineering, Social Justice, and Peace.* His research spans professional and educational reform efforts in engineering, interdisciplinary collaboration in technology design, and the politics of expertise. In addition to two articles in the *International Journal of Engineering, Social Justice, and Peace*, one with Ethan Blue, he has published in *Engineering Studies, Design Studies, Technology and Society,* and *Sustainability: Science, Policy, Practice* as well as several book chapters.